CROP PROTECTION AND SUSTAINABLE AGRICULTURE

The Ciba Foundation is an international scientific and educational charity (Registered Charity No. 313574). It was established in 1947 by the Swiss chemical and pharmaceutical company of CIBA Limited —now Ciba-Geigy Limited. The Foundation operates independently in London under English trust law.

The Ciba Foundation exists to promote international cooperation in biological, medical and chemical research. It organizes about eight international multidisciplinary symposia each year on topics that seem ready for discussion by a small group of research workers. The papers and discussions are published in the Ciba Foundation symposium series. The Foundation also holds many shorter meetings (not published), organized by the Foundation itself or by outside scientific organizations. The staff always welcome suggestions for future meetings.

The Foundation's house at 41 Portland Place, London W1N 4BN, provides facilities for meetings of all kinds. Its Media Resource Service supplies information to journalists on all scientific and technological topics. The library, open five days a week to any graduate in science or medicine, also provides information on scientific meetings throughout the world and answers general enquiries on biomedical and chemical subjects. Scientists from any part of the world may stay in the house during working visits to London.

CROP PROTECTION AND SUSTAINABLE AGRICULTURE

A Wiley-Interscience Publication

1993

JOHN WILEY & SONS

Chichester · New York · Brisbane · Toronto · Singapore

Published in 1993 by John Wiley & Sons Ltd
Baffins Lane, Chichester
West Sussex PO19 1UD, England

Other Wiley Editorial Offices

John Wiley & Sons, Inc., 605 Third Avenue,
New York, NY 10158-0012, USA

Jacaranda Wiley Ltd, G.P.O. Box 859, Brisbane,
Queensland 4001, Australia

John Wiley & Sons (Canada) Ltd, 22 Worcester Road,
Rexdale, Ontario M9W 1L1, Canada

John Wiley & Sons (SEA) Pte Ltd, 37 Jalan Pemimpin #05-04,
Block B, Union Industrial Building, Singapore 2057

Suggested series entry for library catalogues:
Ciba Foundation Symposia

Ciba Foundation Symposium 177
x + 285 pages, 26 figures, 33 tables

Library of Congress Cataloging-in-Publication Data
Crop protection and sustainable agriculture.
    p.   cm.—(Ciba Foundation symposium ; 177)
   Editors: Derek J. Chadwick (organizer) and Joan Marsh.
   Symposium on World Food Production by Means of Sustainable
Agriculture: the Role of Crop Protection, held 30 Nov. to 2 Dec.
1992 in Madras, India.
   "A Wiley–Interscience publication."
   Includes bibliographical references and index.
   ISBN 0 471 93944 7
   1. Plants, Protection of—Congresses.   2. Sustainable agriculture—
Congresses.   3. Food supply—Congresses.   I. Chadwick, Derek.
II. Marsh, Joan.   III. Symposium on World Food Production by Means
of Sustainable Agriculture: the Role of Crop Protection (1992:
Madras, India).   IV. Series.
SB950.A2C76   1993
338.1′62—dc20   93-22791
                                                                            CIP

British Library Cataloguing in Publication Data
A catalogue record for this book is
available from the British Library

ISBN 0 471 93944 7

Phototypeset by Dobbie Typesetting Limited, Tavistock, Devon.
Printed and bound in Great Britain by Biddles Ltd, Guildford.

Contents

v

Participants

K. K. Bhattacharyya Agriculture, Man and Ecology Programme, PO Box 11, Pondicherry 605 001, India

M. M. Escalada Department of Development Communication, Visayas State College of Agriculture, Baybay, Leyte 6521-A, the Philippines

V. Hoon Centre for Research on Sustainable Agricultural & Rural Development, 3rd Cross Street, Taramani Institutional Area, Madras 600 113, India

S. Jayaraj Tamil Nadu Agricultural University, Coimbatore, Tamil Nadu 641 003, India

M. J. Jeger Natural Resources Institute, Central Avenue, Chatham Maritime, Chatham, Kent ME4 4TB, UK

M. J. Jones Farm Resource Management Program, International Center for Agricultural Research in the Dry Areas (ICARDA), PO Box 5466, Aleppo, Syria

P. E. Kenmore United Nations Food & Agriculture Organization, PO Box 1864, Manila, the Philippines

K. A. Lakshmi Department of Botany, Andhra University, Visakhapatnam, Andhra Pradesh 530 003, India

K. Mehrotra Department of Entomology, Institute of Agricultural Sciences, Banaras Hindu University, Varanasi 221 005, India

R. Mishra Maharashtra Hybrid Seeds Company Limited, 4th Floor, Resham Bhawan, 78 Veer Nariman Road, Bombay 400 020, India

S. Nagarajan Indian Council of Agricultural Research, Krishi Bhawan, Dr Rajendra Prasad Road, New Delhi 110 001, India

P. Neuenschwander   International Institute of Tropical Agriculture, BP 080932, Cotonou, Bénin

G. A. Norton   Cooperative Research Centre for Tropical Pest Management, University of Queensland, Brisbane, Queensland 4072, Australia

A. B. Othman   Extension Training & Development Centre, Department of Agriculture, Telok Chengai, 06600 Kuala Kedah, Kedah, Malaysia

M. Pal   Gloria Land, Sri Aurobindo Ashram, Pondicherry 605 002, India

R. Rabbinge   Department of Theoretical Production Ecology, Wageningen Agricultural University, PO Box 430, NL-6700 AK Wageningen, The Netherlands

V. Ragunathan   Central Insecticides Laboratory, Directorate of Plant Protection, Quarantine & Storage, Ministry of Agriculture, 409 B Wing, Sastry Bhavan, New Delhi 110 001, India

D. J. Royle   Department of Agricultural Sciences, University of Bristol, AFRC Institute of Arable Crops Research, Long Ashton Research Station, Long Ashton, Bristol BS18 9AF, UK

S. Savary   ORSTOM, Institut Français de Recherche Scientifique pour le Développment en Coopération, Centre ORSTOM de Montpellier, 911 avenue Agropolis, 34032 Montpellier Cedex, France

R. C. Saxena   International Centre of Insect Physiology & Ecology, Mbita Point Field Station, PO Box 30, Mbita, South Nyanza, Kenya

M. S. Swaminathan   Centre for Research on Sustainable Agricultural & Rural Development, 3rd Cross Street, Taramani Institutional Area, Madras 600 113, India

M. Upton   Department of Agricultural Economics and Management, University of Reading, 4 Earley Gate, Whiteknights Road, PO Box 237, Reading RG6 2AR, UK

A. Varma   Advanced Centre for Plant Virology, Division of Mycology and Plant Pathology, Indian Agricultural Research Institute, New Delhi 110 012, India

K. von Grebmer Plant Protection Division, CIBA-GEIGY AG, PO Box, CH-4002 Basle, Switzerland

H. Waibel Institut für Agrarökonomie, Georg-August-Universität, Platz der Göttinger Sieben 5, D-3400 Göttingen, Germany

J. A. Wightman International Crops Research Institute for the Semi-Arid Tropics, Patancheru, Andhra Pradesh 502 324, India

J. C. Zadoks Department of Phytopathology, Wageningen Agricultural University, PO Box 8025, NL-6700 EE Wageningen, The Netherlands

S. Zeng* Department of Plant Protection, Beijing Agricultural University, Beijing 100094, China

---

*Unfortunately, owing to illness, Professor Zeng was unable to attend the symposium.

Preface

In the autumn of 1992, the Ciba Foundation organized a symposium and open meeting in Madras on World food production by means of sustainable agriculture: the role of crop protection. Our choosing Madras was the result of a very cordial collaboration with the distinguished Indian agricultural scientist, Professor M. S. Swaminathan, who, using funds derived from the first World Food Prize awarded to him in 1987, has established the M. S. Swaminathan Research Foundation and whose Centre for Research on Sustainable Agricultural and Rural Development has recently opened in Madras. The Research Foundation's major aims are 'to integrate the principles of ecological sustainability with those of economic efficiency and social equity in the development and dissemination of farm technologies, to undertake the blending of traditional and frontier technologies in such a manner that opportunities for skilled jobs in the farm and non-farm sectors improve in rural areas, and to develop and introduce technology, knowledge and input delivery and management systems which will enable disadvantaged sections of rural communities, particularly women, to derive full benefit from technological progress'.

It gives me considerable pleasure to record here the great help and support given to the Ciba Foundation by Professor Swaminathan and his colleagues, particularly Dr V. Balaji, during the planning and execution of both meetings. I am also most grateful to Dr G. Thyagarajan, the Director of of the Central Leather Research Institute, for agreeing to host an open meeting.

I very much hope that this record of the wide-ranging papers presented at the symposium and the discussion they stimulated will help to draw wider attention to the crucial issues involved in crop protection and agricultural sustainability and to the important work being undertaken by the M. S. Swaminathan Research Foundation.

Derek J. Chadwick
Director, The Ciba Foundation

Introduction

Jan C. Zadoks

Department of Phytopathology, Wageningen Agricultural University, PO Box 8025, NL-6700 EE Wageningen, The Netherlands

The topic of this symposium is world food production: feeding the growing human population is an obvious task and we don't need to discuss the necessity of doing this. The second part of the title is more interesting: sustainable agriculture. The feeling that we humans are destroying our environment and jeopardizing the future of our own race has gradually crept up on us. Recently, this issue was put in the political limelight by the Brundtland report, which introduced the concept of sustainable development and gave it political credibility.

So sustainability is the issue we are discussing. Following Professor R. Rabbinge, we may see three types of factors at work in agriculture. First there are the production-determining factors, such as sunlight and temperature: with these we might consider the impact of global climatic change and the greenhouse effect. Then there are the production-limiting factors, for example the lack of water or the increased salinity after irrigation. The third group comprises the production-reducing factors. This brings us to crop protection, because pests and diseases reduce food production considerably. The issues of crop protection are the subject of intense debate and great changes in crop protection are pending. It is really a public and political issue in many countries all over the world, tropical and temperate.

The progress of the symposium is rather from the general to the specific. In the first session, we will briefly sketch the ecological background of agriculture. Then we will talk about crop protection: the general aspects and particular economic aspects as a background. The second day will be spent on specific issues and case studies, especially in relation to our host country, India. On the final day we will hear about the success story of India in feeding its growing population from Professor Swaminathan, who is one of the architects of that success.

The ecological background of food production

R. Rabbinge

Department of Theoretical Production Ecology, Wageningen Agricultural University, PO Box 430, NL-6700 AK Wageningen, The Netherlands

Abstract. In the industrialized countries dramatic decreases in the number of people employed in agriculture have been made possible by a rise in soil and labour productivity. There is scope for these to improve further, particularly in developing countries. Potential yields are determined by the characteristics of the crop, local temperature and sunlight. Because the availability of nutrients and that of water are limiting for at least part of the growing season in most agricultural lands, attainable yields are lower than potential yields. Proper management of nutrient inputs, such that optimum use is made of each, can reduce this gap without causing negative environmental side-effects. Actual yields are lower than attainable yields because of growth-reducing factors, such as pests, diseases and weeds. For sustainable agriculture these should be controlled mainly by biological measures. There are many possibilities for this, thus biocides may be used as a last resort not as preventive insurance. Potential yields of rice and sugarcane can reach 30 000 kg ha^{-1} per year of consumable organic matter, sufficient to feed 120 people. Such yields cannot be achieved on all agricultural land, but it is estimated that world food production could support a population of 80 thousand million, if they were all vegetarian and required only 1500 m^2 for non-food-related purposes. The green revolutions that occurred in the Western industrialized countries in the late 1940s and early 1950s and in Asia in the late 1960s and early 1970s need to be followed by a similar increase in agricultural productivity in Africa and West Asia to feed their rapidly growing populations. Better use of fertilizers and good water management require well-educated farmers with the financial means to implement long-term strategies. If these developments are managed properly, food production for the ever-increasing human population can be guaranteed and the burden on the environment and natural habitats reduced, enabling the development of sustainable agricultural systems.

1993 Crop protection and sustainable agriculture. Wiley, Chichester (Ciba Foundation Symposium 177) p 2–29

For centuries, food production was the primary occupation of the majority of the population, and in most countries in the world this remains true. The situation is changing rapidly; for example, until 1860 in The Netherlands more than 50% of the working population was engaged in agriculture, today it is

2

only 5%. In other industrialized countries this percentage is even lower, because *non-soil-dependent* agriculture and horticulture (greenhouses, mushroom cultivation, intensive livestock farming) are more developed in The Netherlands than elsewhere.

The enormous changes in the number of people employed in agriculture have been caused by a rise in soil and labour productivity due to the use of products developed by industry. Investments in land reclamation, mechanization, improvement of soil fertility and crop protection are possible only if industry produces the machines, the fertilizers and the crop protection technologies and agents required. There is scope for soil and labour productivity to increase in 99% of the world's agricultural areas. If these developments are managed properly, food production for the ever-increasing population can be guaranteed and the burden on the environment and natural habitats reduced, enabling the development of sustainable agricultural systems.

History of agriculture

In the early middle ages, French farmers produced some 800 kg of grain per hectare each year of which, because of its poor quality and competition with weeds, 200 kg were needed as seed for the next year's crop. The low level of mechanization meant that it took at least 500 hours to cultivate each hectare. North-western Europe's highly productive agriculture now produces around 7500 kg of grain per hectare per year. No more than 150 kg of seed and 15 hours of work are required for each hectare.

The cause of the low yields per hectare in early medieval times was not the climatic conditions but the chronic shortage of nutrients. Natural fertilization provides only 25 kg of nutrients to plants (nitrogen, phosphate and potassium). In combination with solar energy, this is just enough to produce 1500 kg of biomass, of which 50% is stems, leaves, etc.

Certain agricultural practices, such as the spreading of animal manure or the use of green manure (e.g. clover), increased production. The most important source of plant nutrients was animal manure and the main purpose of keeping large herds on uncultivated land was to improve soil fertility. By the beginning of the 20th century, yields rose to 2000 kg per hectare per year on well-managed land, to which large amounts of animal manure and/or green manure were applied. The small population enabled land to be used in this way in Europe. Enough rangeland was available on which cattle could be kept and thereby enhance soil fertility in concentrated areas where, for example, grain was grown. In other parts of the world, such as China and India, the pressure on agricultural land was much greater and the level of production remained at around 1000 kg of grain per harvest.

In industrializing north-western Europe, population growth increased dramatically in the 19th century and it became impossible to feed everyone on the traditional diet of meat and grain. The introduction of the potato and the replacement of animal fats by vegetable fats enabled many more people to be fed. In potato 80% of the dry matter formed ends up in the harvestable product, the tuber, while for cereals this is only 50%. Crop rotation was employed on a large scale and food crops were alternated with clovers, grasses and other crops used for animal feed. Manure was carefully stored and urban waste was composted and used to improve soil fertility. In other parts of Europe, yields remained extremely low, less than 1000 kg of grain per hectare.

Agriculture with fertilizers

Large-scale increases in yield became possible only after industry began to produce fertilizers. In 1840 the German chemist Liebig showed that plants require only water, minerals and nitrogen from the soil. Organic matter has no nutritional significance, but affects the structure and texture of the soil. Liebig's experiments showed that only 25 kg of the nutrient nitrogen is available to the plant if no fertilizer is applied. By applying minerals to the soil, the level of nutrients, and therefore also yield, could be raised. It was not until several decades after Liebig's discovery that farmers became aware of it and industry began to produce mineral fertilizers. Conditions changed rapidly after that. Up to 1900 productivity rose at a rate of some 3–4 kg of dry matter per hectare per year, after which the rate increased to approximately 15 kg per hectare per year until, after the Second World War, productivity experienced a dramatic increase (see Fig. 1).

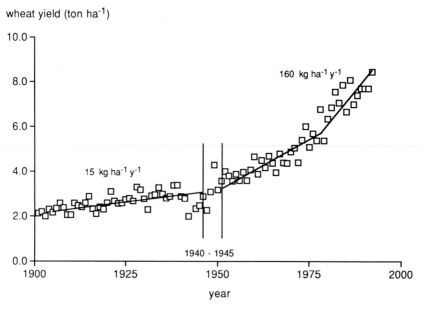

FIG. 1.   Wheat yields in The Netherlands.

TABLE 1 Characteristics of old and new varieties of winter wheat grown under optimal conditions at Cambridge, UK 1984–1986

Variety group	Total above-ground dry matter (ton ha^{-1})	Grain yield (85% dry matter) (ton ha^{-1})	Harvest index (on basis of above-ground dry matter)	Stem length (cm)
Very old	15.0	5.94	0.34	145
Old	15.4	6.55	0.36	134
Intermediate	14.8	7.87	0.45	96
Modern	15.9	9.47	0.51	78

Taken from Austin et al (1989).

Current wheat yields in The Netherlands are approximately 8000 kg per hectare—five times the level at the beginning of this century and twelve times the level in the middle ages. The figures for productivity per unit labour are even more striking—this is now 200 times higher than in the middle ages. This dramatic rise in productivity has enabled large numbers of workers to leave agriculture, and was possible partly because enough jobs were created in other sectors of the economy to absorb the displaced labour. The current high yields are the result of the 200 kg of nutrients applied to each hectare every year, in addition to the nutrients available to plants from the manure produced by animals which during the winter are fed on silage grass and maize, and imported feed. This increase in yield through the use of mineral fertilizers was possible only thanks to the introduction of new varieties, which began in 1840. Some crops, such as buckwheat, were not adapted and play no important role in food provision. Among the cereal crops, the varieties which tiller (develop secondary shoots) readily were gradually replaced by varieties which produce stiff straw (Table 1).

However, the rise in yield over the last hundred years was limited not so much by technical possibilities, as by the demand for food and the ability of the population to pay for it. Since the mid-19th century, farmers have been able to produce all the food required, but only over the last thirty years, at least in Western Europe, the USA and some countries in South America, has income become distributed in such a way that everyone can afford to buy the food they need. This is not true for many countries in Asia which experienced a structural food shortage in the early 1950s. Since the late 1960s, the food production situation has improved considerably, but it is questionable whether this remains so.

Potential, attainable and actual production levels

Potential production

In the industrialized countries, the yield of agricultural crops is still rising rapidly (Fig. 1), despite already being at a high level. This increase cannot continue indefinitely, however, and the question arises: what maximum yields can be achieved, using good varieties, provided with sufficient minerals, nitrogen and water? Total production of organic matter under such conditions depends on the rate of photosynthesis in the green leaves of the crop, expressed as kg CO_2 per hectare per hour. In a single leaf, photosynthesis is directly proportional to light intensity at lower intensities, but at higher light intensities the rate of photosynthesis reaches an upper limit (Fig. 2).

Before light saturation is reached, the slope of the curve does not vary significantly among plant species. The production of sugars at low light intensities is around 0.3 kg ha^{-1} per hour for every Joule absorbed by 1 m^2 of leaf per second. However, the maximum rate of photosynthesis does vary significantly among species. There are two types of photosynthesis: C_4 and C_3 photosynthesis, named after the number of carbon atoms in the first molecule formed after fixation of atmospheric CO_2. The majority of plants have C_3 photosynthesis; only a few (several tropical grasses and crops such as maize and sugarcane) have the C_4 type. At high light intensities, photosynthesis is lower in C_3 plants. This is mainly the result of photorespiration which in C_3 crops increases in proportion to light intensity, but does not occur in C_4 crops. The average maximum rate of photosynthesis in the individual leaves of many of the important agricultural crops, legumes and trees is approximately 20 kg of sugars ha^{-1} per hour. Some tropical crops produce yields at least double these at favourable temperatures.

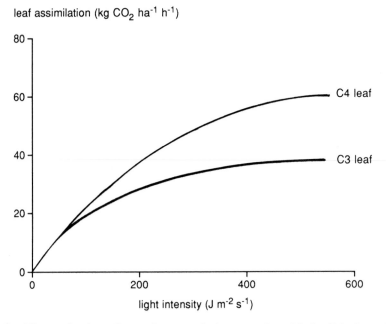

FIG. 2.   The production of organic matter in leaves varies with the light intensity.

At light intensities of approximately 100 J per m^2 per second the leaves of C$_3$ crops have virtually reached their maximum rate of photosynthesis. Such intensities are reached on cloudy days when the sun stands at its zenith; on clear days, intensities of up to 1000 J per m^2 per second may occur. A high proportion of the light reaching crops with one or more layers of large, horizontally positioned leaves is lost. Many crops have narrow leaves arranged in various positions so that light can penetrate deeper into the crop and is therefore distributed more evenly over the leaves.

de Wit (1972) calculated that on a completely clear day closed crops (in which virtually all light is intercepted) whose individual leaves have a maximum photosynthetic rate of 20 kg of sugars ha^{-1} per hour and a leaf arrangement like that of a cereal crop, photosynthesize at rates of 35, 50 and 55 kg of sugars ha^{-1} per hour when the sun stands at an angle of 30°, 60° and 90° above the horizon, respectively. When the sky is cloudy, light intensities are approximately one fifth of those found under clear conditions. However, the rate of photosynthesis is reduced by no more than one half because light is distributed more evenly throughout the crop under cloudy conditions and because photosynthesis uses proportionally more of the light energy at lower light intensities. In The Netherlands, potential photosynthesis is approximately 400 kg of sugars ha^{-1} per day in summer and 200 kg ha^{-1} per day in spring and autumn. If the daily totals from mid-April to mid-October are added, potential photosynthesis of a healthy crop surface is approximately 50 000 kg of sugars ha^{-1}.

The sugars produced are not stored as such, but are used by the plant to produce its roots, stems, leaves, flowers, fruits and seeds. The production of 1 g of proteins, fats or cellulose and absorption of 1 g of minerals requires 1.92, 3.23, 1.28 and 0.12 g of sugar, respectively. For 1 g of plant material containing 25% protein, 5% fat, 60% cellulose and 10% minerals by weight, 1.42 g of sugar is needed. This means that the rate of 400 kg of sugars ha^{-1} per day which is possible in June produces a plant growth of 275 kg of organic matter ha^{-1} per day. Roughly one quarter of this organic matter is used in respiration. Therefore estimates of potential yield must assume production of 200 kg of dry matter ha^{-1} per day. This growth rate actually occurs in The Netherlands under favourable conditions (Fig. 3).

Attainable and actual production

Potential yield is achieved only under exceptional conditions. Usually, the attainable yield is considerably lower, because for part or even all of the growing season, growth is restricted by shortage of water and/or nutrients. In addition, crops are plagued by diseases, pests and weeds. In the majority of the world's agricultural regions, the attainable yield is less than 20% of the potential yield. The actual yield is generally even lower, because agricultural practices are not carried out optimally.

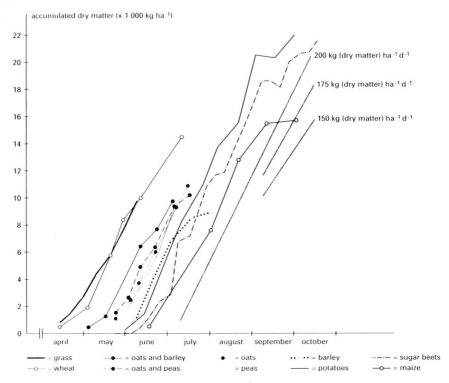

accumulated dry matter (x 1 000 kg ha⁻¹)

FIG. 3. Growth rates of different crops in The Netherlands throughout the growing season. The growth curves correspond fairly closely to the theoretical production rate of 200 kg of dry matter ha^{-1} per day. From Sibma (1968).

In areas with a high level of land reclamation, which is the case for most of the world's agricultural land, actual yield is often further below attainable yield than in areas with a low level of land reclamation. The difference between high and low levels of reclamation can be seen in the degree to which attainable production approximates potential production. Actual yield can be raised to the level of attainable yield by the use of good cultivation practices, particularly in areas with a high land reclamation level. One can thus differentiate between potential, attainable and actual yield.

The potential yield is determined by growth-defining factors, i.e. incoming solar radiation and temperature and the characteristics of the crop. These include physiological (photosynthetic) characteristics, its phenological characteristics (crop development), the optical properties of the leaf (reflection, transmission and absorption of radiation) and its geometric characteristics (leaf arrangement and ability to intercept radiation).

The attainable yield is that which can be achieved under conditions of sub-optimal amounts of growth-limiting factors (see Fig. 4). In virtually all the

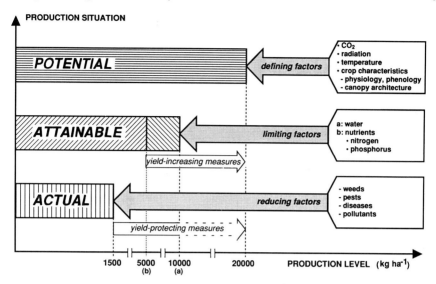

FIG. 4. The relationships among potential, attainable and actual yields and growth-defining, growth-limiting and growth-reducing factors.

agricultural areas of the world, growth rate and, therefore, production are limited during at least a part of the growing season by a shortage of water and nutrients. An analysis of growth and production of Sahelian rangelands showed that for part of the growth period phosphate is in short supply, then nitrogen; at other times shortage of water is the limiting factor. If the problem of poor soil fertility were eliminated, the attainable yield would be 2–5 times the present yield. Despite appearances, water is not the sole growth-limiting factor in this case.

Factors which reduce growth include diseases, pests and weeds, and these also reduce the attainable yield. The prevailing weeds, pests and diseases at low production levels are different from those at high production levels. Crop growth-limiting and crop growth-reducing factors interact. Quantification of the effect of these factors on the underlying processes of light interception, assimilation, respiration and transpiration and the effect on, for example, water and nutrient uptake and integration in models of crop growth, helps in understanding the various growth-reducing factors and provides a basis for the development of control measures.

Weeds compete with crops for environmental resources that are in limited supply. Mechanical and chemical weeding methods may be applied. Timing and appropriate control methods depend on detailed knowledge of the consequences of weed occurrence and activity during various development stages (Kropff et al 1984). Pests and diseases affect crop growth at all production levels, but are important at only some.

The impact of pests and diseases on crop performance depends on the mode of interaction, which is determined by the morphological and physiological characteristics of both crop and the pest or disease agent, and their interaction with environmental conditions. A similar approach may be followed for pollutants, when their point of impact in plant processes is unravelled. The effects of pests and diseases can be incorporated into a crop growth simulation model, and thereby quantified. Some of these interactions will be exemplified for a number of pests and diseases.

Crop growth-reducing factors

Stand reducers.   Organisms such as damping-off fungi or caterpillars in massive numbers negatively affect plant density. This may decrease yield unless remaining plants have the ability to compensate for these losses by increased tillering, branching or locally increasing leaf area.

Light stealers.   The assimilation rate may be reduced by a direct effect of the pathogen on light use efficiency through 'light stealing'. Some leaf pathogens (e.g. perthotrophic and saprophytic fungi) cause the death of host tissue, which may remain and absorb radiation without assimilation taking place and reduce light penetration into lower leaf layers. The mycelium produced may cover surfaces of living leaves and reduce light absorption or interfere directly with photosynthesis (see assimilation rate reducers, below) through excretion products or necrotic lesions.

Assimilation rate reducers.   Many pathogens decrease the assimilation rate by changing the physiological properties of the leaf. Viruses (e.g. beet yellows virus on sugar beet) and some fungi may affect the photochemical and biochemical processes in chloroplasts (reflected in lower values of both A_{max} and ϵ) or reduce the number of chloroplasts per unit leaf area. Bacteria may cause structural damage to chloroplasts. Some pathogens and insects may accelerate leaf senescence. Other pathogens such as rusts may perturb functioning of the stomata, resulting in higher resistance to CO_2 uptake and hence a lower assimilation rate.

Assimilate sappers and tissue consumers.   Assimilate sappers feed on primary assimilates (e.g. aphids and mites drain assimilates from the parenchymal cells), while tissue consumers remove crop tissue (e.g. cereal leaf beetles). Since each kg of glucose produces less than 1 kg of structural dry matter, tissue consumers are more detrimental in terms of crop growth per unit weight consumed. However, secondary effects of assimilate consumers (e.g. interaction with plant physiological processes, attraction of other pathogens), may result in greater total growth reduction.

Some pathogens affect the turgor of cells, either directly by suction or by disrupting the tissue such that transpiration increases. The water balance of a plant may also be affected by root pathogens and nematodes that feed on the roots. These may also disturb the crop nutrient balance by interfering with phloem transport to the roots, thereby reducing energy availability for active uptake of nutrients, and by disrupting the passive flow of water and nutrients because of decay of conducting tissue.

Pollutants. Chemicals such as heavy metals and gases may reduce crop production as they reach toxic levels. At this point, plant physiology may be influenced through competition (nutrient uptake, gas exchange) or through modification of environmental conditions at cell level (osmotic strength, acidity).

Control of pests and diseases

Because pest and disease infestations are fundamentally different from weeds, in terms of mobility and interactions with plants, control operations have to be adapted accordingly. Timeliness of the control measures is crucial.

Preventive measures against pathogens include removal and destruction of crop residues to reduce contamination potential; crop rotations including fallowing to reduce nematode and other soil pathogen populations (e.g. in greenhouses); crop breeding for resistance against pathogens. However, in many cases pathogens break crop resistance after some time, so that the breeder has a 'race without an end'.

Preventive measures against pests include the use of repellents, pheromones (to attract enemies), nets for some insects and birds or fencing for animals. Breeding in this context primarily aims at modifying morphological characteristics, such as taller spikelets in wheat to hamper bird attacks. Such modifications may unfavourably affect harvesting procedures.

Curative control is based largely on the use of biocides, but both pathogens and pests may develop resistance against chemicals. Biological control agents (such as wasps, killing larvae or whitefly) are widely studied and tested. In designing proper techniques for these control measures, modelling of population dynamics is an important tool.

Finally, technology is applied to protect the harvested products. Preventive measures include drying, cooling, heating, acidification and fumigation, some of which can also be used curatively.

Potential production in the world

As indicated earlier, potential photosynthesis in The Netherlands is around 50 000 kg of sugars or 35 000 kg of organic matter ha^{-1} per year. Given that part of this is needed for respiration and that only part of the biomass is suitable for human consumption, we can assume that, with our current level of knowledge and the crop varieties available, it should, in a commercial setting, be possible to obtain 35% of the biomass in a form suitable for human consumption. There are 14 MJ per kg of dry matter so this comes to 175 GJ ha^{-1} per year. Since one person requires 3.5 GJ of energy each year, in The Netherlands one hectare could feed around fifty people a year, assuming that they ate only vegetable products and did not waste much food. In other parts of the world potential production differs, depending on the length of the growing season, the temperature and the level of solar radiation. If enough nutrients and water are supplied, it is possible to grow crops all year round in the tropics. This gives a potential photosynthesis of some 120 000 kg of sugars ha^{-1} per year, or a production of approximately 30 000 kg of organic matter suitable for consumption, which could feed 120 people for a year. These yields have been shown to be attainable with crops such as rice and sugarcane.

These calculations demonstrate that the world population, which currently stands at over five thousand million persons, could, under European conditions, feed itself from an area of 100 million ha. The current amount of agricultural land in the European Community is 127 million ha. However, by no means all of this land is capable of producing such high yields. Less than 60% of the cultivated land in Europe would be capable of reaching such levels of production. Nevertheless, the potential for food production is unimaginably big. Naturally, a situation in which Europe grew food to feed the rest of the world is highly unlikely, but it demonstrates the enormous potential of agricultural production. In fact, this theoretical situation illustrates the upper limits of cultivation practices, although it does not take into account the socio-economic and ecological constraints.

If potential production is achieved, 150 m^2 is required for food for one person and more than 1×10^{12} people could live from the total land area (excluding the oceans) (Table 2). This figure is more than 200 times the current world population. However, this number of people could live from the earth, but would not have enough space to live on the earth.

Estimates of the amount of land one person needs for food, work and relaxation depend strongly on the cultural background of the person making the estimate. If we assume that at least five times the area needed for food is needed for other purposes, we obtain 750 m^2 per person. This is still a small area and has been calculated rather arbitrarily.

On the basis of this figure and an average of 150 m^2 for food production, a total of 900 m^2 per person is required. This allows for a maximum world population of 145×10^9. In The Netherlands, 15 million people live on some 30 000 km^2, approximately 20 000 km^2 of which is used for agriculture and horticulture. This population should require only 20% of the land area for its own food production, rather than the current 60%. This land, suitable for modern agricultural practices, is available.

TABLE 2 The potential production of all land area between 70° N and 50° S and the potential number of people who could be fed

Degrees North	Total land area (10⁸ ha)	Months with with temp > 10 °C	Total organic matter (1000 kg ha⁻¹ per year) (potential production)	Area required (m²) per person (after conversion to useful production)	No. of people (10⁹)
70	8	1	12	806	10
60	14	2	21	469	30
50	16	6	59	169	95
40	15	9	91	110	136
30	17	11	113	89	151
20	13	12	124	81	105
10	10	12	124	81	77
0	14	12	116	86	121
− 10	7	12	117	85	87
− 20	9	12	123	81	112
− 30	7	12	121	83	88
− 40	1	8	89	113	9
− 50	1	1	12	833	1
Total	131				1022

However, humans do not live solely by potatoes; if meat is to be included on the menu, approximately twice the land area is required, assuming that only a small amount of meat figures in the diet, since it takes 10 kg of vegetable matter to produce 1 kg of meat. This reduces the number of people which Earth can support to

$$\frac{150 + 750}{(2 \times 150) + 750} \times 145 = 125 \times 10^9$$

This is still an impressive figure.

However, there will be those who believe that each person requires 1500 m² to live on and for non-food-related purposes. The maximum number of people who could live on Earth, when excluding meat consumption, would then be

$$\frac{150 + 750}{150 + 1500} \times 145 = 80 \times 10^9$$

The figures demonstrate that the size of the world's population depends not so much on the land area required for food production, but on the land desired for other purposes. To a certain extent, therefore, scope for food production need not restrict the size of the population. Irritation is likely to be a more important factor, along with the need to dispose of waste products.

de Wit made the above estimates in 1975, on the basis of a very simplified calculation. Since then, more accurate estimates have been obtained for various regions using computer simulation models. These models use properties of the soil and climate as basic data and simulate the growth and production of different crops (quantitative analysis). The location, the prevailing climate and the properties of the crop are taken into account, then the potential and attainable production, based on the availability of growth factors, can be calculated. A comprehensive analysis of the potentials of crop growth and food production in various regions of the world, based on soil and climatic conditions was done by Buringh et al (1975). On behalf of the European Community, fifteen years later the Netherlands Scientific Council for Government Policy (WRR) commissioned the Winand Staring Centre (SC-DLO) in Wageningen to carry out an even more detailed analysis. All their estimates show that, on the basis of the potential soil productivity, production levels several times the current ones could be obtained. If the land area suitable for different types of agriculture is determined (qualitative analysis), it becomes clear, for instance, that the land area in Greece which is suitable for arable farming (the most demanding type of land use) is only 10% of the total area. In the other 90%, the land is too steep (making mechanized agriculture impossible), or the soil is too shallow, too saline, too acidic or too rocky. In other countries, such as Denmark and The Netherlands, more than 50% of the total area is suitable for demanding forms of land use. Through a combination of quantitative and qualitative land evaluation, a fairly accurate assessment can be made of the potential and attainable yields for different crops under the different conditions found throughout the world. The analysis of SC-DLO was used by the WRR for a study of possible developments in the agricultural regions of the European Community. This study indicated that agricultural production could be several times higher than it is at present. Nevertheless, European agriculture, certainly in the most suitable areas, is more productive than in the rest of the industrialized world and much more productive than in developing countries (Netherlands Scientific Council for Government Policy 1992).

Production efficiency: green revolution

Labour productivity and production efficiency

The rise in labour and soil productivity mentioned above is expected to continue for some time, for two reasons.

First, potential production is much higher than actual production. In 99% of the world's agricultural areas, soil productivity is considerably lower than its potential. Worldwide, less than 15% of potential production is achieved. The most important cause of increasing soil productivity resulting from innovation lies in the efficiency of use of inputs. It will be demonstrated below that, at higher production levels in good production situations, efficiency in terms of input per unit product is higher than at lower production levels. The effect of this increase in efficiency at higher production levels promotes continuing growth in production per unit area.

The rise in soil productivity over the past few decades has been accompanied not only by an increased efficiency in input use, but also by higher labour productivity. Around the turn of the century, the production of one tonne of wheat in The Netherlands required 300 hours of labour, the same amount can now be obtained with about 1.5 hours of labour. The growth in the demand side of the market has been another significant incentive fostering growth in production.

The additional inputs needed to achieve these higher yields are sometimes grouped together on the basis of their energy content (Table 3). In contrast to prevailing intuition and many energy surveys conducted in the 1970s, Table 3 shows that the fully mechanized, high-yielding American maize-growing industry is three times more efficient in its use of energy than are traditional methods of cultivation where all work is carried out by hand or by animal traction and where no industrial fertilizers are used. Merely grouping together all energy inputs,

TABLE 3   Energy production and consumption related to four different methods of maize production

	Nitrogen consumption (kg ha^{-1})	Yield (kg ha^{-1})	Output GJ ha^{-1}	Input GJ ha^{-1}	Output/ input
Mexico Only human labour, no industrial fertilizers	0	1944	28.89	39.40	0.73
Mexico Human labour and oxen, no industrial fertilizers	0	941	13.98	19.26	0.72
USA Human labour, horses, industrial fertilizers	152	7000	102.58	111.79	0.92
USA Human labour, machines, industrial fertilizers	152	7000	102.58	48.2	2.14

Input calculated as all energy inputs (direct and indirect). Data from Pimentel & Hall (1984).

including food and fuel, is not always useful, nevertheless it shows something. On the basis of direct and indirect energy consumption, this comparison of production techniques shows that the Law of Diminishing Returns, which is valid for single inputs under constant conditions, does not hold for agricultural production as a whole. This is because energy is not an input with a single effect, but a resource which can be used in varying amounts depending on the production level and degree of technological development. The increasing efficiency of energy use demonstrates that technological advances in agriculture make possible increasing yields with a relatively lower unit of input per unit of product.

The reason for this is that the relative costs of fixed activities in agriculture, such as ploughing and sowing, decrease as yield increases. In principle, a farmer does not need to plough or sow more to obtain higher yields. For instance, to obtain modest yields, the acidity of the soil has to be adjusted by applying lime, but higher yields do not require more lime than lower yields. The same applies to plant nutrients. This means that many inputs are not variable costs, but are determined by the decision to grow a particular crop. The number of activities needed to grow a crop in a given production situation requires a set of fixed costs; others, such as application of fertilizer or pest and disease control, can be done in various ways and involve variable costs. The better the production situation through structural improvement, the higher are the fixed costs relative to the variable costs. Therefore the variable costs decrease as yields rise. No farmer will improve water management on his land without improving soil fertility by using fertilizer. He will also ensure that this crop is adequately protected. In areas with a low level of land reclamation, crop protection and fertilizers can be regarded as variable costs, because the efficiency of inputs is generally low. In areas with a high level of land reclamation, they are considered essential and therefore are fixed costs. This statement challenges the widespread prejudice that variation in inputs completely depends on their prices. This does not hold for the primary inputs where substitution possibilities are very limited, but does for the secondary inputs where substitution is possible. For example, weeding may be done mechanically or with pesticides, so that labour and capital may substitute for each other.

One consequence of this increase in fixed activities over variable activities is that at higher yields the applied inputs can be better controlled than at lower yields. In situations where high yields are obtained, growth processes are better understood and managed. In low-yielding situations, the effects of various external factors are subject to a stronger mutual influence. Application of the energy-demanding nitrogen fertilizers is better controlled in high-yielding situations, where the unforeseeable losses resulting from volatilization, denitrification, leaching and immobilization are greatly reduced, because of better water management and soil structure.

Changes in production situations due to structural agricultural measures, such as improved land reclamation levels, may lead to higher attainable yields. Concomitantly, the yield at which the application efficiency of external inputs, in terms of their use per unit of product, is highest rises. In well endowed regions (good soils, high land reclamation levels, i.e. plentiful water and nitrogen), the production levels at which optimum use of external inputs is realized are high (de Wit 1992). In regions with poor soils and low levels of land reclamation, the production level at which the optimum use of external inputs is achieved is generally low.

At low land reclamation levels, high production may be attained by using very high inputs. The application efficiency per unit of product is in those cases generally very low. Improvement of land reclamation through structural changes in the physical production conditions will increase the efficiency of use of external inputs. Poorly endowed regions or marginal lands are agriculturally unproductive and in environmental terms dangerous. It is hazardous to use them.

Wheat yields in The Netherlands rose from 3500 kg ha^{-1} in 1950 to 7500 kg ha^{-1} in 1990, while the output relative to input of direct and indirect energy rose from 145 kg of wheat per GJ to more than 200 kg of wheat per GJ. The increase in labour productivity and the concomitant increase in energy input were four- or fivefold over the same period.

Crop protection is also an important precondition for higher yields. It does not take much energy to apply the correct biological, mechanical and chemical crop protection methods, but it does require knowledge and experience. The farmer must therefore be highly skilled. Lack of skill can be compensated for by applying excessive chemical crop protection agents; however, that is bad agricultural practice. Integrated protection against diseases and pests, in which as much use as possible is made of their natural enemies and the crops' resistance, together with preventive phytosanitary measures, requires frequent field observations. By using suitable measures at the right time and in the correct way, much loss of production can be prevented. The same applies to other agricultural practices.

High production levels do not, therefore, necessarily require more chemical energy in the form of fertilizers, machines and pesticides; rather they call for well-trained farmers who are capable of taking well-considered decisions throughout the growing period of a crop. Brain power is much more effective than energy in the form of tangible inputs, and the amount of energy it takes to think is negligible.

Yields continue to increase until the attainable level is reached, which for a high level of land reclamation lies only just below the potential yield (Fig. 4). The increase in yield per unit area is determined not so much by economic factors as by the rate at which knowledge and experience are assimiliated and put into practice by the farmer. If it is economically feasible to farm, suitable resources and technology should be properly used. In this way of thinking, it is not a matter of whether to use more or less technology or inputs, but whether one is going to farm or not.

Green revolutions

The rise in labour and soil productivity discussed above has been accompanied by a number of breaks with tradition. In the late 1940s and early 1950s there was a sudden increase in the growth of grain production per unit area in the Western industrialized world (Fig. 1). This first green revolution, as it was known, was the result of a combination of developments in several scientific disciplines. Short-stem varieties of plants, which had been bred by Heine during the Second World War, were introduced; the use of nitrogen fertilizers increased rapidly; herbicides were introduced. The resulting rise in yield and the rapid mechanization of farm work raised labour productivity to unprecedented levels.

This first green revolution (which went largely unnoticed by the public) was followed in the late 1960s and early 1970s by a second green revolution in Asia, particularly India, China and Indonesia. The increase in productivity that occurred there ended the structural food shortages which had plagued that region since the early 1950s, despite the growing population.

Many parts of the world, particularly Africa and parts of the Middle East, now need a third green revolution, because they have never experienced the kind of explosion in productivity described above and have burgeoning populations.

Environmental effects of agriculture

Environmental effects of poverty

For centuries humans have had to face the problems which arise when land is not used properly. The depletion of soils and overuse of irrigation systems have caused erosion and the irreversible loss of good agricultural land. The bare hills around the Mediterranean, particularly in Greece, bear witness to this tragedy. Sand drift in The Netherlands was caused by human activity. The over-exploitation of the natural environment by the Aborigines in Australia made large areas of land unsuitable for agriculture.

Until the green revolutions, the harmful effects of agriculture on the environment and natural habitats were the result of the over-exploitation of the potential of agro-ecosystems. This still constitutes the main threat to the majority of the world's agricultural regions. Over-exploitation of this kind is not confined to developing countries. Large-scale agriculture—which virtually amounts to overcropping—in Australia, the USA, South America and the Commonwealth of Independent States could seriously threaten the continuity of agriculture there. The dust storms which occurred in the USA in the 1930s could now happen again as a consequence of cavalier environmental management. Agricultural systems geared merely to short-term economic gains can have a drastic effect on the environment and threaten agriculture in the long term. Erosion is a particularly serious threat. Agricultural methods which ensure that soil does not become depleted, that layers of soil do not wash or blow away and that the structure and texture of the soil are kept intact are available, but according to current economic thinking are often unattractive.

A good farmer ensures the continuity of his farm, but can do so only if the right preconditions are created and there is no net economic gain to be had from exhaustive cultivation. The threat to the world's food supply as a result of underutilizing inputs and overly extensive agriculture is enormous.

A typical example is found in Africa. Sub-Saharan Africa is the only part of the world where food production per capita has declined over the past two decades as a result of an increasing population on one hand and lower yields on the other. The latter has, among other things, been attributed to catastrophes such as drought and infestations of locusts. A more gradually developing problem in the region, less readily linked to food shortages of the recent past, is declining soil fertility.

To calculate nutrient balances, Stoorvogel & Smaling (1990) partitioned the arable land of 38 countries in Africa into agro-ecological zones, characterized by rainfall, current soil fertility level, cropping systems, application of fertilizer and manure, crop residue management and erosion control. The flow of nutrients into and out of the soil was assumed to be governed by five input and five output factors.

The net balance of inputs and outputs of nutrients was negative in all countries considered (Fig. 5). The highest annual nutrient depletion rates were found in densely populated East Africa. West Africa had moderate to high depletion rates; Central Africa and the Sahel region had moderate to low rates. Soils in the Sahel region are so poor that little is left to lose.

The highest depletion rates are in most cases associated with erosion of the top soil rather than export in crop products. The difference between inputs and outputs is covered by the soil nutrient reserves: with progressive depletion, the soil nutrient supply decreases and crop production will decline. On the whole, nitrogen deficiency is almost universal in Africa and fertilizer application is a prerequisite for crop production. However, because of interactions with other nutrients, application of nitrogenous fertilizers only leads to more efficient 'mining' of other soil nutrients. Nutrient deficiencies can be corrected only by integrated use of organic and inorganic sources of various nutrients.

Environmental effects of wealth

This underutilization of inputs in large areas of the world is in sharp contrast to the overutilization prevalent in Western Europe. A surplus of animal manure and the virtually constant price of fertilizers over the last 15 years in Western

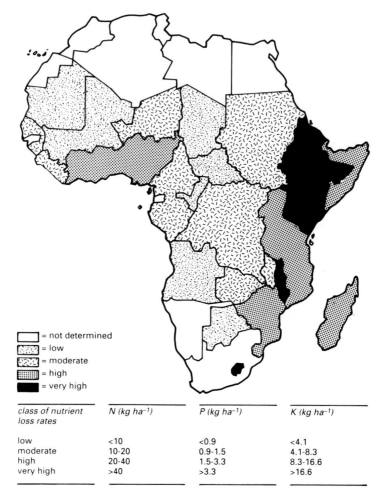

class of nutrient loss rates	N (kg ha⁻¹)	P (kg ha⁻¹)	K (kg ha⁻¹)
low	<10	<0.9	<4.1
moderate	10-20	0.9-1.5	4.1-8.3
high	20-40	1.5-3.3	8.3-16.6
very high	>40	>3.3	>16.6

FIG. 5.   Rates of nutrient loss from arable land in Sub-Saharan Africa in 1983. From Stoorvogel & Smaling (1990).

Europe, which has been a consequence of improved production processes, has meant that the relative price of fertilizers is now so low that farmers are encouraged to overuse it. In The Netherlands the excessive use of nitrogen from animal manure and fertilizers has reached drastic proportions. An average of 550 kg of nitrogen is applied to each of The Netherlands' 1.1×10^6 ha of grassland each year in the form of fertilizers and animal manure; only 75 kg of nitrogen from each hectare finds its way into the milk and meat produced. Large amounts of nitrogen therefore accumulate in the environment.

The discrepancy between nitrogen input and output is caused partly by the unclear relationship between the technically and economically optimum levels of input of nitrogen. The maxim that 'what doesn't do any good can't do any harm' has been applied in the case of nitrogen, but is untrue. Technically, no benefit is gained from the overuse of nitrogen: it does not result in higher yields. It produces indirect economic benefits through feed intake by animals. However, that can be attained technically in a much better way through changes in diets. The implicit gap between the technical optimum and the economic optimum would thus seem to be one cause of this overutilization. However, because overuse of nitrogen fertilizers is of limited economic benefit, the gap between these optima cannot be the only reason for this practice. In simple terms, overuse can result in 'peace of mind' for the farmer: the thought that, at any rate, the crop has sufficient nitrogen. The constraints on farmers' finances these days prompt many of them to adopt this attitude, to combine maximum yields with minimum risks.

The best way of stopping the overuse of nitrogen fertilizers is to give farmers better information about fertilizing their land, the link between fertilizers, crop growth and the environmental impact of overuse of fertilizers. This will not in itself be enough: a financial incentive in the form of a tax on nitrogen fertilizers is required, on theoretical macro-economic as well as practical grounds. This would reduce the actual use of nitrogen fertilizers to approach the technical optimum (generally 200 kg of nitrogen ha^{-1} per year for grassland). Another environmentally favourable development is the shift in agricultural production towards more productive land, where less nitrogen leaches into groundwater and much higher nitrogen output in products is achieved per unit of input. Harmful environmental effects resulting from prosperity can therefore be solved by concentrating production in smaller areas, within integrated agricultural systems, which also frees areas for natural development.

Detrimental environmental effects resulting from the underutilization of inputs can be tackled only if more use is made of external production factors (mainly fertilizers and good water management). Particularly in areas of low soil fertility, care must be taken to restrict expansion of the area under cultivation. Expansion could lead to use of marginal land, quickly causing soil depletion, after which the land is abandoned. Uncontrolled expansion of land under cultivation— caused by rising populations—entails huge environmental risks, which can be averted only if a rise in productivity can be achieved in fertile areas, within limits which ensure that the environment is protected.

Primary and secondary production

So far, this paper has been devoted almost exclusively to crop production. Animal production is derived from crop production. As a result of increasing prosperity and therefore increasing demand for animal products, there has been a sharp rise in the demand for animal feed. This generally requires 4–10 times more agricultural land than production of vegetable crops that are directly consumed by humans. Thus, a growing demand for meat results in more land being used to grow crops. The rise in productivity per unit area discussed above implies that this increase has been limited.

In the past, crops for animal consumption were produced on the farm where the animals were kept. This is still the case in the majority of the world's agricultural systems. However, in The Netherlands, the existence of the port of Rotterdam and its specialization in the import of cheap animal feed, and the fact that farms are too small to produce food for the large number of animals kept on them, has meant that the link between animal and vegetable production has to a large extent been broken. The crops required to feed livestock are produced elsewhere in the world, while animal production—mostly in the form of dairy products and meat—takes place in The Netherlands. A significant proportion of animal products is exported, while the nitrogen in the manure produced remains behind, creating huge environmental problems. Restoring the link between animal and crop production would go some way towards solving these. This would require both technical measures, such as improvements in the composition of animal feed, the transport and processing of manure, and a reduction in the number of livestock. This 'luxury' problem is not intractable: the means exist, it is merely a question of finding the political will.

Acknowledgements

I would like to thank Bjørn Dirks and Harrie Lövenstein for their assistance in the preparation of this paper. A great deal of the information is taken from de Wit (1971).

References

Austin RB, Ford MA, Morgan CL 1989 Genetic improvement in the yield of winter wheat: a further evaluation. J Agric Sci 112:295–301
Buringh P, Van Heemst HDJ, Staring GJ 1975 Computation of the absolute maximum food production of the world. Wageningen Agricultural University, Wageningen
de Wit CT 1971 Voedselproduktie: verleden, heden en toekomst. Stikstof 6:396–408
de Wit CT 1972 Food production: past, present and future. Stikstof 15:65–80
de Wit CT 1992 Resource use efficiency in agriculture. Agric Syst 40:125–151
Kropff MJ, Vossen FJH, Spitters CJT, de Groot W 1984 Competition between a maize crop and a natural population of *Echinochloa crus-gall* (L.). Neth J Agri Sci 32:324–327 (synop)
Netherlands Scientific Council for Government Policy 1992 Ground for choices. Four perspectives for the rural areas in the European Community. Sdu Publishers, The Hague (Rep Gov 42)
Pimentel D, Hall CW 1984 Food and energy resources. Academic Press, New York
Sibma L 1968 Growth of closed green crop surfaces in The Netherlands. Neth J Agric Sci 16:211–216
Stoorvogel JJ, Smaling EMA 1990 Assessment of the soil nutrient depletion in Sub-Saharan Africa. Report 28. Winand Staring Centre, Wageningen

DISCUSSION

Kenmore: One pest we will be hearing a lot about in the next few days with reference to rice is the brown plant hopper. This eats and excretes sugar solution from the phloem of rice plants—about 1 mg of dry matter per day. So it would take about one million plant hoppers on one hectare to eat the equivalent of 1 kg. One million plant hoppers is four per plant at normal planting densities, which is less than the commonly observed level. The question is what happens to the crop yield? Does the plant hopper really impose any limit to production? I have a feeling it's just another sink: the crop will respond, over a fairly wide range of pest densities, to that additional sink.

Rabbinge: If you calculate the amount of assimilates consumed by a particular insect, in general terms it's always very low, compared with the plant growth that potentially can be achieved. If the insect is just consuming leaf matter or dry matter, in many cases it is just another sink to the crop that, in situations where sinks are limiting crop growth, may promote production. Leaf folders in rice are a typical example. If the leaf folder is not present at the early stages of plant development, limitation of sink size may limit development of the source. So in some cases, an insect or a disease actually raises production. Many crops growing under optimal conditions are not source limited, they are sink limited.

Secondly, many of the pests, in addition to the direct effect of consuming leaf mass, have an indirect effect. For example, aphids promote early senescence; they also reduce the efficiency of light use and photosynthesis. These effects are much more severe than the direct effect of consuming assimilates and may cause damage at a much lower level of insect infestation than one would predict just on the basis of the energy balance. Thus, in some cases, it is better to have 'pests' than to be without; however, in other cases, even a limited infection may cause heavy losses.

Kenmore: If we are trying to sustain a productive ecosystem in a field, an indigenous second trophic level (herbivores) may be needed to keep the third trophic level (predators) happy and dense enough to deal with the occasional pulse of immigrant herbivores.

Rabbinge: Yes, a crop that is completely free of insects or herbivores is a sterile environment. In such situations, effects due to sink limitations are present. The plant may therefore not fully use the incoming radiation for growth and production.

Varma: Another important indirect effect of insects like aphids, leaf hoppers and whiteflies is the transmission of viral diseases.

For direct effects, Marion Watson (1967) worked out that if you left aphids unchecked, they would suck off 100 lb of sap per day from each acre of sugar beet crop. Fortunately, that does not happen in any system because ecological balance is maintained through predators and parasites.

Jayaraj: There are two other issues concerning herbivory by sucking insects. One is the phytotoxaemia caused by certain insects that introduce toxins along with their saliva. Secondly, there is the removal of plant auxins along with the plant sap. It is not merely the defaecation that counts, these factors are also important.

I would like to comment on four inputs—nutrients, water, crop variety and crop protection materials. In many countries that are poor in natural resources, you commonly find wrong use of fertilizers. The non-availability of fertilizers at the right time is a major constraint. The main problem is that if farmers don't get financial credit on time, they are unable to purchase the fertilizers. So the limiting nutrient, namely nitrogen, is applied when it can be obtained, which is often not at the recommended times. This has resulted in increased pest and disease activity, particularly in rice—tungro virus and leaf folder. We need to promote more integrated nutrient management for sustainability, including greater use of organic manure, like green manure and biofertilizers.

The second input is water. Very often, Asian farmers inundate rice fields to suppress weeds, but this predisposes the crop to attack by the brown plant hopper and other pests. These occur at high densities at the critical crop growth phases and, again, dry matter production is greatly affected.

Thirdly, crop variety: any production situation is governed by the interaction among three factors—genotype, crop management and the environment. Crop management varies according to the resource base of the farmer. Recently, the environment has become completely distorted, so the optimum dry matter production of the genotype is not obtained.

Lastly, there is considerable wastage of chemical pesticide input owing to incorrect formulation, the wrong method of application and so on. This may also be true with the new biopesticides, pheromones, etc, that are becoming available. The application equipment should be based on the needs of the local situation. If these points are observed, we can promote sustainability in food production.

Zadoks: The correct timing of agricultural measures is a point that is often overlooked; it seems to me that this is especially so in tropical countries which are adopting modern technology. Part of this is due to the lack of credit at the right time, as you said, and here we enter the realm of social sciences. This is why we have so many social scientists at this meeting; we cannot work towards sustainable crop protection without support from social scientists.

Rabbinge: I support the comment that proper nitrogen fertilization is very important for maximum efficiency. I agree that very often nitrogen is not applied at the right moment. When it is applied at the proper time according to the production situation and the stage of crop development, it has a synergistic effect.

If you plot nitrogen uptake against the application rate, you see that the basic amount of nitrogen available to a plant without any treatment varies according

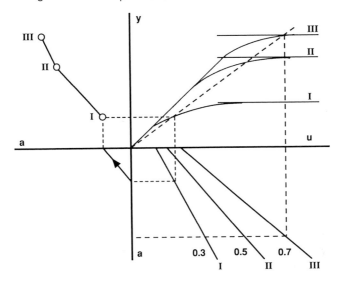

FIG. 1. (*Rabbinge*) The effect of various levels of land reclamation (I, II and III) on the relation between fertilizer uptake (u) and yield (y), on the relation between fertilizer application rate (a) and uptake, and on the relation between fertilizer application rate and yield.

to the level of land reclamation (Fig. 1). At a better level of land reclamation, the level of basic nitrogen available at which an equilibrium between inputs and outputs is reached is higher. The slope of this line, which reflects the efficiency of use of nitrogen, also changes. At a better level of land reclamation, the efficiency increases. If you have proper control of pests and diseases at the right times, you see a similar shift—an increasing basic amount of nitrogen available to the plant and better use of the nitrogen. So by treating with nitrogen in the right amount and at the right time, you can increase efficiency synergistically. This is a very important point. It means that enough of these basic elements have to be available, therefore it's very important to apply the fertilizer at the right moment. In many countries in the developing world, shortage of nitrogen is the problem. So there is a continuing danger of poverty, because the nitrogen or the fertilizer or other input is not available at the right time. On the other hand, for example in some parts of Western Europe, there is tremendous overuse of external inputs, such as nutrients or pesticides, which is hazardous for the continuity of the whole system and is very unsustainable.

Kenmore: We are talking about sustainability, not in an idealistic sense of a closed system, but as sustainable *growth in food production*, without which we cannot open our mouths in Asian policy-making circles. We must sustain growth, not just sustain ecosystem flow. Basically, building up the productive parts of the ecosystem—which for insect control means predators, the third trophic level—early in the season and keeping them available, is not only the cheapest but also the most reliable response to crop phasing, where there are critical periods of yield vulnerability to stress, including fluctuations in the populations of herbivores. Evolution gives us a stage, as it were, to play on, and we need to maximize what we can do on it. This means building up predator densities to levels that are routinely attainable and can handle, in the case of many rice insects, a change in herbivore density of an order of magnitude in 3–4 days.

Varma: I am very happy that Dr Rabbinge expressed the belief that we can feed a global population of 80 thousand million people. That provides grounds for optimism for countries with growing populations. Let's hope it is possible.

You mentioned that in the United States there has been an increase in production of wheat per unit area. But Ruttan (1987) claimed that in the United States the increase in production per unit worker was greater than that per unit area between 1880 and 1980. During the same period in countries like Japan and Germany, where the development started with a relatively high population:land ratio, the increase in output per unit land has been largely responsible for the growth in total productivity. In contrast, in India, production per unit labour has declined slightly (Varma 1989). Recently, in parts of India, a decrease in production of food grains has been observed in areas where high inputs are used (Varma & Sinha 1992). Have you come across such a decrease in yield under European situations?

Rabbinge: We do not have that situation in Europe as far as I know. For the majority of crops, yields are still increasing—that's creating problems for the European Community, because we have a surplus of production. One reason for this increase in productivity, especially in southern parts of Europe, is that there is still a big gap between potential yields and actual yields. In many cases, they are producing only 20–30% of the potential; whereas in parts of Great Britain, Schleswig-Holstein in northern Germany and in The Netherlands, they are producing 80–90% of the potential.

In India, the stagnation of yields may be due to the fact that the various external inputs are probably not applied in the proper combinations.

Swaminathan: I am glad this was the opening paper, because in India and other developing countries, our greatest concern in agricultural research is how to achieve as much of the potential as possible, without damaging the environment. Professor Varma mentioned yield stagnation, for example in the Punjab. Already the farmers are producing 10–12 t ha^{-1} of rice and wheat, with a harvest index of about 40% (i.e. 40% of the energy from total photosynthesis goes to grain formation). Sugarcane farmers in India are almost realizing the full yield potential: in parts of Tamil Nadu state nearly 90% of the production potential has been achieved. The major problem is: at what cost can it be sustained?

The Chinese increased their total production of food grains from 300 million tonnes to 400 million tonnes between 1980 and 1990 (World Bank 1991, unpublished). They would like to increase it by another 100 million tonnes between now and the year 2000. But the increase to 400 million tonnes was achieved largely through the use of additional mineral fertilizer and there is now nitrate pollution of the ground water. In India, the areas where sugarcane is being produced at very high yields (200 to 300 t ha^{-1}) are all areas where ground water is being depleted very rapidly.

So the challenge before this meeting is: under conditions where we must achieve as much of the potential production as possible, how do we do it such that we don't jeopardize the prospects of long-term production? In India, almost 50% of the land area is under agriculture; the area of forest has dwindled enormously, land is a shrinking resource for farming. Under these conditions, the only way to increase overall production is to reach the potential production level, but what are the implications? In Europe production is rising, but they have high prices and very high subsidies—the economic aspects of productivity improvement are important. Are there clear studies, in The Netherlands for example, on the ecological implications of this very high level of achievement in terms of potential production?

Rabbinge: In many situations in The Netherlands, sustainability has been endangered by the overuse of external inputs for two reasons. First, the crop rotations are too narrow. Potato always gives a better economic result than do small grains, so frequently potato occurs too often in the rotation. This creates difficulties and necessitates fumigation of the soil to prevent an upsurge of nematodes and other pests. If we return to broader crop rotations, there will be a considerable reduction in pesticide use.

Secondly, there is incorrect use of pesticides. If these were used in the proper way in the European Community, we have calculated that annual pesticide use could be reduced from 400 million kg of active ingredients to between 40 and 80 million kg. This could be achieved in three ways. (1) Use of the best land at high production levels with very high efficiency and low emissions. (2) Adoption of crop rotations that prevent systematic overuse of pesticides and better integration of crop and animal production. (3) Application of good housekeeping methods, such as integrated pest and disease management.

The same is true for nitrogen. There has been tremendous overuse of nitrogen because it is very cheap. The actual price of nitrogen during the last 20 years has not increased, so the relative price has decreased considerably. Now we are saying, we should have broader crop rotations and more intelligent use of fertilizers. Many farmers have already decreased the nitrogen input. For example, in potato and sugar beet, during the last 20 years, nitrogen input has been reduced by more than 50% but yields have still increased, because of better use of external resources.

So there is certainly a way to increase production in India and other developing countries, but it needs a lot more knowledge of how to manipulate the crop—not only in the heads of research workers, but in the farmer's fields, which of course is the difficult thing.

Royle: All this high input doesn't necessarily mean that we are very effective at controlling pests and diseases in Europe. Husbandry methods have inevitably changed during intensification, which has created new problems. There is evidence for shifts in the dominance of some pests and diseases. Surveys have indicated that relative to changes in production levels over the last 20 years, we are no better at controlling certain diseases now than we were then, even with heavy pesticide use. This makes a scenario which can be difficult to modify towards sustainability, and in which it will be difficult to find methods that are appropriate for sustainability.

Jayaraj: The input efficiency varies from temperate to tropical situations. In the tropics, the loss of fertilizer nitrogen, due to run-off, seepage and volatilization, occurs much faster. The organic carbon content is degraded rapidly because of the high temperatures and light intensities, which are not found in temperate climates. So sustainability is more at stake in the tropics than in the temperate regions. We may have to develop technology to compensate for this.

In this context, Dr Rabbinge's comment that ecotechnological insight and maximal biological self reliance are vital for development is highly significant. In India recently, Dr Swaminathan has introduced the biovillage concept to promote everything biological to sustain agriculture. This seems to be one answer, to promote biological self-reliance.

Lastly, methane emission, increasing UV irradiation and the effects of other greenhouse gases will all be serious concerns with regard to sustainability. Studies have been started by the Indian Council of Agricultural Research and in many other countries on the effects of these greenhouse gases on plants. We should also observe the effects on pests and diseases, on the expression of host plant resistance to pests and diseases, on the natural enemies and on artificially introduced biological control agents.

Saxena: Concerning high input agriculture, plants, particularly green plants are intrinsically resistant to a certain degree of herbivory. But if they are stressed by nutrient imbalance, their innate defences break down. We determined the comparative allelochemical fractions in rice plants. Production of these chemicals per unit plant was lower in plants grown under stressful conditions than in plants grown under conditions of proper nutrient balance. Several other abiotic stresses, e.g. salinity, iron toxicity, aluminium toxicity and high temperature, have the same effect. Sustainability could be improved by the use of cultivars that are tolerant to these abiotic stresses.

Rabbinge: Nutrient imbalance in a crop creates difficulties, not only because of direct stress, but also because it often promotes the rapid development of particular pests and diseases. If there is overuse of nitrogen, then there are many free amino acids in many of these crops, which favours a stronger upsurge of, for example, aphids. So it's necessary to minimize the use of nitrogen in such a way that you maximize the efficiency of use of the phosphorus and potassium. It's always this mini-max relation, among all these external inputs. At the optimum levels, when you have the proper combination of these external inputs, then you have a synergistic effect of all the different inputs, including crop protection measures—and that is our ultimate aim.

References

Ruttan VW 1987 Agricultural research policy and development. FAO, Rome (FAO Res Technol Pap 2)

Varma A, Sinha SK 1992 Sustainable development through long-term biotechnological alternatives in agriculture. In: Daniel RR, Ravichandran V (eds) Proceedings of the International Seminar on Impacts of Biotechnology in Agriculture & Food in Developing Countries. Committee on Science and Technology in Developing Countries (COSTED), Madras, p 44–57

Watson MA 1967 Epidemiology of aphid-transmitted plant virus diseases. Outlook Agric 5:155–166

Sustainable agriculture: an explanation of a concept

M. J. Jones

Farm Resource Management Program, International Center for Agricultural Research in the Dry Areas (ICARDA), PO Box 5466, Aleppo, Syria

Abstract. Close relationships between agriculture and many areas of human activity determine countless interlinkages with global issues of natural environment protection, human population increase, food supply, industry and world trade. This broad context promotes different perceptions of sustainable agriculture by different interest groups. Profitable diversification away from overproduction of basic commodities and satisfaction of environmental pressure groups are major preoccupations in developed countries. Elsewhere the main concern is to maintain a trend of increasing production: food security with a future dimension. Achieving this depends essentially on protecting the agricultural resource base. Inputs and input substitution are important co-related issues but the core of sustainability is the avoidance of any attrition of the *potential* for future production; this demands that we guard soil, water sources, grazing lands and gene pools against loss and degradation. Though superficially biophysical or technical in nature, most problems of resource degradation and eroding potential are rooted in economic, social and political issues; few such problems will be solved unless the primacy of these issues is recognized and addressed. Sustainable agriculture will likely remain elusive until governments and other agencies accept it as arising only as the outcome of a synthesis of strategies on population, employment, economic planning, technical research and national investment.

1993 Crop protection and sustainable agriculture. Wiley, Chichester (Ciba Foundation Symposium 177) p 30–47

Use of the word 'sustainability' has become very fashionable in recent years. As anxieties about mankind's impact on the biosphere have proliferated, 'sustainability' has been increasingly used to express a desirable goal in such global spheres of concern as the environment (conservation of habitats and biodiversity *versus* exploitation, degradation and pollution); food security; security of fuel supplies and industrial feedstocks; human equity, current *and* intergenerational; population growth; and world trade and political order. For each of these, sustainability has the connotation of maintaining or improving (in some sense) the *status quo*.

Agriculture is a major participant in all of these spheres. It threatens the environment with habitat loss, degradation and pollution; it provides the food; it uses fuel and industrially produced inputs (but also, increasingly, produces fuel and other materials needed by industry); it threatens (through current abuse) the resource base for future generations; and it provides occupation/livelihood for a large proportion of the world's population, thereby promoting political stability and supplying the materials for world trade. Any concept of agricultural sustainability must accommodate these dynamic interactions, positive and negative, between agricultural activities and environment and society.

Sustainability in world agriculture

Because of the breadth of the context, and because of its ramifications into nearly all areas of human activity, agricultural sustainability is perceived very differently by different interest groups. Tinker referred to the 'two cultures': that relating to the putatively overstressed and overproductive agricultural systems of Europe and the USA, and that of Third World agriculture, for which the perceived context is shortage of food and land (P. B. Tinker, unpublished paper, RASE Int Symp Towards Sust Crop Prod Syst, Cambridge, July 1992). For the agriculture of the developed world, the main sustainability issues are diversification away from a limited range of commodities and the satisfaction of environmental pressure groups, particularly in respect of the large flows of nutrients and pesticides currently used. For developing countries the imperative is to maintain food production (or, more realistically, to maintain a trend of increasing food production), while preserving (and enhancing) the underlying resource base.

Although our principal concern here is with the developing world, it is worth reviewing briefly what is happening in the developed world as it seeks its own brand of agricultural sustainability. What happens there will have its impact on farmers everywhere. Driving forces for change in European and North American agriculture include: recent overproduction of conventional crops; resulting changes in the European Community's Common Agricultural Policy and the consequent release of land for new enterprises; the increasing domination of agriculture by commercial and industrial interests, the first demanding products that meet stringent 'quality' requirements for efficient mass marketing, the second seeking cheaper and ever more various and specific feedstocks for chemical and biochemical processes.

In this context, for what are increasingly becoming highly capitalized systems, sustainability tends to mean mainly economic sustainability. The avoidance of pollution is a major issue, but great interest also centres on new and prospective technologies, aimed at increasing the range and value of agricultural products. This has several dangers for the Third World. One is the increasing ability of biotechnology to facilitate the production in temperate areas of: crops previously limited to warmer latitudes; new crops, whose products may at least partly substitute for those previously imported; and crop varieties specifically tailored for certain pesticides, which may have little relevance in the Third World. At the same time, developing countries may find it increasingly difficult, when marketing their agricultural products, to meet ever tighter quality specifications for horticultural crops and the required chemical/biochemical compositions of bulk commodities. The developed economies are industrializing their agriculture, increasing its range, and increasing their capability to supply their own needs. Meanwhile, continuing overproduction of basic commodities remains a political disincentive to assist developing countries to become self-sufficient.

Sustainability in developing countries

First, it is important to be sure what it is we wish to sustain. Some authorities see sustainability simply as an extension into the future of current concerns about food security: if production is maintained (or, better, increased) then we have sustainability. But is that adequate? The maintenance of production is a necessary index of sustainability, but is it sufficient? Production is output from farming systems utilizing available resources, and we need to consider what is happening to those factors also.

In respect of farming systems, opinions vary: for some, the concept of sustainability includes the preservation (with improvements) of existing farming systems, to sustain social stability, livelihoods, local cultures, etc. This view has some force. We cannot ignore quality of life and human welfare, but nor can we expect to preserve systems which, whether biophysically/ecologically sustainable or not, no longer match the wider realities of economics and politics. We seek sustainable systems but not necessarily system sustainability. At low population densities, slash-and-burn agriculture is arguably an indefinitely sustainable system in tropical forests, but there can be few areas for which this would be the recommended form of land use for the twenty-first century and beyond. New, equally sustainable but more productive utilization is required. Realistically, we must expect all farming systems to be dynamic, to evolve or be replaced over time, as technology changes and, particularly, as human needs and perceptions of need change. Sustainability is not concerned with resisting this.

Defence of the agricultural resource base is a different matter. If the resources supporting agriculture are allowed to diminish or deteriorate, production becomes less easily sustainable, irrespective of the farming (or social or political) system. The bottom line of all debate on sustainability is the obligation to safeguard the potential for future production. Resources is another word that tends to mean different things to different people. In fact, the resources utilized for agricultural production are of two different general types: (1) internal resources: natural resources, like soil, water and native vegetation at the production locations. (2) External resources: materials brought in from outside, like diesel, fertilizer and agricultural chemicals.

External resources

Current production levels in many high-yielding farming systems depend on the utilization of externally derived inputs and many of the prospects for production improvements in lower-yielding systems are predicated on increased inputs. From the standpoint of sustainability, there are dangers in this. External inputs may make the utilization of fragile and marginal natural resources temporarily profitable (e.g. fertilizer use on shallow, sloping soils; cheap fuel for overpumping aquifers and cultivating marginally productive land) while simultaneously promoting degradation. Further, most externally derived inputs are based on global resources of finite size, in particular fossil fuels. Long-term continuity of cheap supplies cannot be assured. A major requirement for sustainability is therefore the minimization, wherever possible, of the dependence of production on such inputs.

This last statement needs clarification, since it is often misunderstood. For most parts of the developing world it is unrealistic to suggest that we should attempt to discourage, in the cause of 'sustainability', the reasonable use of fertilizer, pesticides and farm machinery. At the same time, however, our research and development effort should be seeking economically viable substitutes, full and partial, for those inputs—through research in such fields as integrated pest management, reduced tillage, supplemental irrigation, biological nitrogen fixation, nutrient recycling and efficient use of fertilizer. We should also be monitoring the impacts of inputs on the natural resource base, both locally and in the broader environment. (Pollution and public health, though not yet such public issues as they are in developed countries, cannot be disregarded.)

Internal resources

Whatever the contribution from external resources, the essential foundation of agricultural sustainability everywhere is the protection, maintenance and enhancement of the natural resource base, the soil, water and vegetation that constitute the internal resources of every farming system. Sustainability is about the conservation of the productive capacity of those resources. For example: crop production potential declines, inevitably, if arable soils are allowed to degrade and erode; irrigation cannot continue indefinitely, if the aquifers that supply it are utilized more rapidly than they are replenished; livestock production from rangeland cannot be maintained if the palatable plant species are eradicated by overgrazing; and, biotechnology notwithstanding, the general potential of agriculture to meet future demands is diminished, if biodiversity and the gene pools of crops, animals and range and pasture plants are reduced.

Few people disagree with such propositions. Nevertheless, with the urgency of current needs and the practical limitations imposed by lack of knowledge, poverty and imperfect human institutions, many compromises are made that diminish future potential. Any hope of improvement lies in a better understanding of the many heterogeneous, and often interacting, issues involved. A rough separation of these issues into three areas may be helpful.

The three Rs: renewability, reversibility, resilience

Resources are sometimes classed as renewable or non-renewable. Water in a perennial river is a renewable resource. Non-renewability is harder to define. Soil is effectively non-renewable: if it is washed away, replacement by natural processes is very slow; *but* under careful management the same soil may be utilized productively indefinitely. In contrast, fossil water is non-renewable *and* can be used only once.

Such distinctions are germane to the evolving economic philosophy of resource utilization. In the face of evident loss of potential owing to consumption and degradation, economists are seeking ways to include such processes in their calculations. The simplest approach is to regard non-renewable resources as capital and production from them as interest. Prudent economic management demands that capital be preserved and, if possible, increased, and current expenditure be financed out of the interest. Living off one's capital is non-sustainable.

In fact, that view makes little sense for resources that can be used only once. Fossil water (like fossil fuel) is capital that bears no interest and there is no persuasive logic against using it now rather than later. In the cause of sustainability, one can only urge: use it slowly, and use the time thereby gained to look for a more permanent replacement for the future.

However, resources like soil and natural ecosystems are capital that can be either squandered or managed to yield a sustainable interest. In the real world, the latter (ideal) situation is always under threat from urgent needs that propose a trade-off between yield now and some loss of subsequent potential. The economic practice of future discounting—whereby the *current* value to an individual (or society) of a particular resource is greater now and diminishes into the future, according to a subjectively judged discount rate—tends to encourage this attitude in development planning.

Minor degradation of potential to meet urgent short-term production needs is sometimes defended on the grounds that it is reversible (the depleted aquifer will refill *if* we stop overpumping for a while; the pasture will recover *if* we exclude the animals for a few seasons; new topsoil can be built up again on the deep subsoil under appropriate management). Parallels are drawn to natural disasters, which inflict 'perturbations' on ecosystems (even in the total absence of humans). The ability of an ecosystem to recover completely from a perturbation is termed resilience; systems of low resilience are described as fragile. These concepts are applicable equally to agricultural or natural ecosystems. The study of agricultural systems from the viewpoint of resource ecology is a rapidly expanding field of technical research necessary to put our understanding of sustainability on a firm scientific basis.

But the time scales of biological, hydrological and soil processes are long, and it can take many years to identify, measure and understand mechanisms of resilience, and to discern whether particular losses of potential are reversible or not. Some damage, even irreversible damage, may arise from very gradual, insidious processes. Meanwhile, for immediate practical purposes, sustainability is best served by minimizing all agricultural perturbations—to a degree commensurate with the evident fragility of the resources, judged subjectively.

Scale and the cultivation frontier

Issues of sustainability arise on all geographic scales: from individual farmer's plots, through village lands and small catchments, to provinces, major river basins and ecoregions. There are 'knock-on' or 'downstream' effects from one area to another and perturbations that are reversible on a small scale (e.g. isolated slash-and-burn plots) may be disastrously degrading and irreversible on a large scale.

Because much of the action must come from governments, national perceptions of society's obligation (or ultimate self-interest) to manage natural resources sustainably, balancing concerns for present and future food security, are crucial. A fact of life in many countries is the colonization by the plough of previously non-arable land: driven by rapid population growth, the cultivation frontier advances further each year into more fragile land. A real dilemma at national (and international) level is whether to put research, development and control effort into saving the fragile areas and their usually rather meagre production potential or to try to relieve the pressure at the frontier by seeking greater output from more favoured, less fragile high-production areas. Some argue that this latter, indirect route is the most effective way to conserve resources and ensure sustainability.

Of course, it is not always as easy as that. Short of draconian countermeasures, people will continue to live and seek their livelihood in fragile areas; and even the production of ample supplies of food elsewhere is no guarantee that degradation of resources will be curtailed. Economic realities often discourage equitable distribution. Moreoever, attempts to increase production in favoured areas may begin to stress even the more resilient resource bases. There are recent indications of plateaux, or even declines, in yield in some of the very productive systems promoted by the green revolution; the reasons underlying this are not understood. The agricultural sustainability of high-production areas cannot be taken for granted. Sustainability is not just a question (as is often portrayed) of protecting the fragile lands and resources. Vigilance is required in all zones and systems, and government efforts have (somehow!) to be allocated in a balanced fashion between them.

People and policies

Agricultural sustainability is sometimes perceived as being mainly a matter of solving technical problems: if we identify, through research, the right technologies and transfer them to the resource users, all will be well. But that is rarely the case. Although some adaptation may be needed locally, the right technologies are often already known and those most appropriate are frequently part of the tradition of the resource-users' society. If such technologies are no longer used, the reasons most likely lie in a complex of local circumstances that may be broadly labelled socioeconomic. For instance, if terraces are not maintained, this is probably because the cost or 'opportunity cost' (the cost of pursuing one option rather than alternatives) of repairing them is seen to be too high, not because no-one knows about terraces. A little enquiry usually shows that where mismanagement of resources is threatening sustainability, the root causes are economic or sociopolitical rather than simply technical.

Certain controversial factors recur through many agricultural systems and societies. One, in particular, is land tenure. Farmers are rarely willing to invest much effort and cash in land in which they do not have long-term rights of ownership or usufruct. Even where ownership is confirmed, the land may consist of scattered patches or a strip running up a hillslope, in either case confounding the individual's best intentions to utilize the soil conservatively.

Common property is another issue. The 'distribution of property rights in resources in which a number of owners are co-equal to use the resources' (R. S. Pomeroy, unpublished paper, Meet CGIAR Soc Sci, ISNAR, The Hague, August 1992) occurs widely in respect of land (particularly grazing land), water and even hunting and fishing territory. Formerly, village elders (or committees) apportioned individual access to new land or to a limited irrigation source; tribal groups each knew their grazing lands, managed them in a traditional (and, therefore, probably sustainable) way and defended them against outsiders. Now, many of these local controls have broken down—the results of population pressure, social disruption and, often, direct central government interference—and the accumulated wisdom associated with them is endangered. Common property systems have been widely replaced by 'open access regimes'; in other words, a free-for-all. People tap into stream and aquifer as they wish, and government regulations (where they exist) have much less force than previous obligations to observe local custom. The loss to central government of the former tribal control of the steppe grazing lands of western Asia has been a major factor in their recent degradation.

Such social, economic and political factors are at the heart of many of the problems of resource management, which threaten sustainability, particularly but not exclusively, in more fragile environments. Prospects for continued productivity at current low levels, not to mention sustained increases, depend on a judicious local blend of reforms, incentives, effective controls and, where appropriate, technical innovation, accepted by the people and guaranteed and supported by both societal and government institutions. As Cernea put it:

'. . . for good environmental management, addressing the environmental requirements *alone* is utterly insufficient: we are under the imperative to address also the *sociological* variables which are paramount for the use of natural resources *by people.*'

(M. M. Cernea, unpublished paper, EDI Semin Nat Res Environ Manage Dry Area, ICARDA, Aleppo, February 1992).

Synthesis

Some of the issues surrounding the concept of sustainable agriculture have now been outlined. For those who would like a succinct definition, I cannot improve on that recently given by Izac et al (A.-M. N. Izac, M. J. Swift & K. A. Dvorak, unpublished paper, Meet CGIAR Soc Sci, ISNAR, The Hague, August 1992). They described sustainable agriculture as a 'complex concept incorporating issues of ecological stability and resilience (e.g. conservation of resources and reduction of impacts on the environment), economic viability and the quality of life and human welfare'.

Undoubtedly, the concept is complex, at least in the variety of its specific applications to so many widely different environments and, in each of them, to a continuum of interlocking technical, economic and political factors. At another level, though, it is remarkably simple: it is a universal imperative to guard (and where possible enhance) the potential for production by current agricultural systems and by all conceivable future agricultural systems. Such an imperative needs to become axiomatic—as an almost unconscious principle—to all agricultural planning, whether that of the national policy maker or of the individual farmer deciding next season's operations. For this, the concept needs to become part of the culture at every level in all agricultural societies.

As already suggested, the concept of sustainability existed in some form, intuitively or as the product of long experience, within the collective wisdom of many traditional farming communities. One much quoted example is 'hema': this communal management system with ancient roots, by which local tribal authority over pastoral lands, in what is now Arabia, Syria and Iraq, afforded resource protection through a complex of migratory, semi-sedentary and deferred grazing practices (T. L. Nordblom, unpublished paper, EDI Semin Nat Res Environ Manage Dry Area, ICARDA, Aleppo, February 1992).

In some situations, such old systems can perhaps be revived and modified to meet modern needs. Elsewhere, new models and new approaches are required to cope with changed conditions. One such model is 'Landcare', an Australian programme for management of natural resources in an agricultural context (D. Marston, unpublished paper, EDI Semin Nat Res Environ Manage Dry Area, ICARDA, Aleppo, February 1992). Of course, the conditions (agricultural, social, economic) are very different in Australia from those of most developing countries, but Landcare stresses community involvement, the appropriate valuation of environmental assets and the creation of a new ethos, involving a change of thinking about the responsibility for and use of the nation's natural resources. Those ideas apply to all societies. The seminal word is ethos—a climate of opinion by which the community (on all geographical scales) refuses to allow individuals to degrade a communal resource, irrespective of nominal ownership. The development of that ethos, in common with the concept of population regulation that relates so closely to agricultural sustainability, is a major educational task—perhaps the most pressing educational task in the world today.

DISCUSSION

Kenmore: This issue of ethos and ethics is important. We will get further if we think of people not as part of the problem but as part of the solution—at village level and all the way up. You can look at that formally, as an economist, in terms of human capital. Another approach is that adopted by Dr P. Pingali at IRRI. Yield ceilings and production ceilings on the IRRI farm itself routinely hit $9\,t\,ha^{-1}$, but the average yields are not increasing. It depends on how you look at the germplasm, but basically average yields have been declining about 1–2% per year for the last 15 years.

What do we do about this? The answer is what are being called second-generation technologies in green revolution environments, that empower farmer decision-making. Management tools are being made available to farmers so that their farming is smarter: the farmers are using resources in more creative ways and are doing better applied research.

I was glad Mike Jones mentioned traditional farming systems. Indigenous knowledge or technology should not be seen as a fossilized entity that belongs in a museum. The indigenous research processes and indigenous resources include the people in the village. They are able to do trials: they can take new germplasm or new kinds of economic systems as new challenges and modify them through local adaptive research, perhaps in partnership with formal research. In India, there are Operations Research Project sites where researchers lived in villages and worked with farmers to develop locally appropriate techniques. Although these are one of the best examples, they need institutional support and strengthening to expand. Training and education are not just a matter of hammering people with normative technical statements; people need tools with which they can analyse and do research in their own agro-ecosystem.

Jones: The human dimension is by far the most important. This is not only for the reasons you have stated; once you get into management of resources like soil and water, the scale of operations becomes so much larger that you can't do research unless you work with the people. The research that I have tried to initiate at ICARDA is based on social science, with technical aspects as an additional feature.

Upton: The farmers are not always the best guardians of the environment and resources. Farmers have been responsible for Syria's soil depletion in some situations.

Jones: I agree. The first step is always to find out why farmers are not the best guardians. What is pressing them to behave in this way? Farmers are not stupid: if they have a time horizon of only one year, they have a good reason. You have to discover that reason before you have a hope of improving their ability to handle natural resources. The initial requirement is diagnosis of resource problems rather than diagnosis of the yield problems.

Upton: I would like to ask about the problem of common resources. Do you think the introduction of individual ownership offers advantages in terms of resource conservation?

Jones: I don't know. Some resources are not amenable to 'privatization'. Aquifers are an example. In Syria at the moment, aquifers are used for irrigation where there is no surface water. The only rational approach of a landholder there is to dig a borehole quickly and extract as much water as possible, because otherwise neighbouring farmers will take all the water. What is the best approach in that kind of situation, I just don't know. We are thinking about it; it is a social science problem.

Kenmore: Recent empirical work on common property has been summarized by Robert Wade (1988). The work was done mostly in Andhra Pradesh. Villagers defined the conditions for sharing water, and Wade described, over a range of natural water availabilities, what kind of common property allocation mechanisms had been worked out in semi-arid zones. (These areas are not as dry as where Mike Jones is working.) The study showed that there are social controls at village level and means for villagers to negotiate access to common properties.

Hoon: We have started a biovillage project in Pondicherry in southern India. Biovillage is a new approach to sustainable agriculture and rural development. The aim is to integrate recent advances in biotechnology with the best in traditional techniques, in a way that enables the livelihood security of rural people to be improved ecologically and economically.

TABLE 1 *(Hoon)* **The impact of technological changes in crop production on soil fertility and the water table in India**

Year	Number of crops	Crop variety	Duration (days)	Yield (t/ha)	Number of wells	Well depth (m)	Soil status
1950–1960	1	Gudreval	170	4	2	10	Excellent
1960–1970	2	IR8	135	6	4	30	Good
1970–1980	3	Jaya/vaigai	125/115	5.6	7	40	Chemical fertilizers and pesticides used
1980–1990	3	ADT36/ES18	105/90	5.2	20	60	Strong dependence on agrochemicals
1991	3	ADT36	105	4.8	31	80	

As part of the our farmers' participatory research programme, I interviewed several farmers to assess the impact of green revolution practices on crop yields, soil health and the depth of the water table. The farmers traced the impact of the green revolution and the changes they had seen over the past 40 years. The results are summarized in Table 1.

I found that there was steady depletion of water resources; the water table is falling at a rate of 1m per year. The farmers are aware of this but they are totally guided by profits. They know that they are not maintaining soil fertility; they know that in 20 years their own children may not be able to farm the land; but it's the immediate future they worry about. They may say: 'My daughter is to be married in three years so I will plant this crop because it will guarantee me a profit. I know that this is may not be the best possible use of the land, but I need the money.' Profit is the main motivating force. Also, there are no mechanisms or incentives for conserving water. Each farmer feels that even if he uses water judiciously, others will not and he will be the loser.

In this biovillage project, our aim is to maintain profits for the farmers by providing economically and ecologically suitable alternatives. Currently, they plant three crops of paddy per year, which draws heavily on the water resources. In the traditional crop rotations, there was at least one dry farming crop, such as millet, but these water and soil conserving practices have been abandoned. Some of them need to be reintroduced. We need cooperation amongst government workers, farmers and NGOs that work with farmers. This is something we have consciously tried to incorporate into our rural development schemes.

Ragunathan: Concerning sustainability at a village level or farm level without disturbance to the environment, farmers have been getting good yields even without the use of chemical pesticides. Many of us have been associated with the promotion of integrated pest management (IPM) at the village level. With the active support and guidance of Peter Kenmore, several demonstration farms, each of 100 ha, have been established, on which farmers and extension workers are being trained in IPM techniques. Two weeks ago, in a sector of the field in the IPM demonstration area, an over-enthusiastic farmer sprayed chemical pesticides for the control of leaf folder where it was not warranted. Because the larvae of this pest remain inside the folded leaves, the spray was ineffective. However, it did cause irreparable damage to the naturally occurring beneficial species, including predators and parasites that control the leaf folder population. As a result, the entire field was damaged by the pest, while the rest of the fields in the IPM demonstration area remained pest free. It is evident from this experience that application of chemical pesticides disturbs the balance of the ecosystem and thereby affects farmers' yields and agricultural sustainability.

Neuenschwander: Mike, there are many instances where agricultural land borders on non-agricultural land. One of the values of the latter is the preservation of species diversity through habitat conservation. If such habitats are used for agriculture, you might be criticized by environmentalists. At the International Institute of Tropical Agriculture we are looking at interactions between farms and uncultivated land. This is particularly important in Africa, where there is a relatively high proportion of untouched land—not virgin rainforest, but relatively untouched land and the actual fields constitute only a small percentage of the surface area. It would have been interesting to have somebody from the environmental side at this symposium. The farmers are not stupid, but the people who defend natural habitats are not stupid either. It would be interesting to contrast those opinions. Agriculture is embedded in a wider environment and on the human side, there is more than just the farmers, there is the whole society.

Wightman: I support what Peter Neuenschwander has just said, referring to Africa. I carried out surveys on groundnut crops on farms in three or four countries in Africa, looking particularly at soil insects, because they were affecting the groundnut crop most. There was a relationship between the size of farm, which can be taken as an indication of the intensity of the farming, and the amount of damage caused. In the intensive farming systems of Malawi and Zimbabwe there was a lot of damage by soil insects. In Zambia, the farms were larger and the farming was less intense, and there was less damage. In Tanzania, where slash-and-burn agriculture is still quite common, insect damage and disease losses were difficult to detect.

From these observations, I've realized that we shouldn't look only at the farmed area. When assessing the potential of various pests to cause damage, it is necessary to look at the effect produced by the farms on the non-urban environment as a whole. The non-farm areas represent reserves of natural enemies. It is these that keep potentially damaging insect pests under control; the smaller the reserve, the less stable the system.

Because of my lack of experience with rice crop, I find it difficult to rationalize this train of thought with what happens in a paddy system, where there is very little area which isn't paddy. But I think this is a very important aspect of sustainability in non-intensive systems.

Kenmore: The species diversity in paddy is staggering. Bert Barrion of IRRI, who is South East Asia's expert on rice arthropod taxonomy, in paddies that are producing $4-7 \text{ t ha}^{-1}$ with no insecticides, routinely finds 400–600 species of arthropods during an average three month season on about 10 t of standing plant biomass. Plant biomass is important for creating niches, like parking spaces in a garage, that the insects and spiders can stand on. In the tropical rainforests, there are about 400 t of standing plant biomass per hectare. If you take Terry Erwin's figures from the Amazon of 5000–8000 species per hectare, maybe as many as 10 000 (although Robert May has questioned some of these estimates), the species diversity of arthropods in a tropical rice paddy, in terms of species per tonne of standing plant biomass, is of the same order of magnitude as in the Amazon rainforest. The resilience and the renewability of tropical systems, even under intensive management, a managed ecosystem that only exists for three and a half months before it's harvested, are orders of magnitude beyond what many people are used to thinking about. There is a tremendous species diversity in paddy that is worth conserving.

Savary: The number of arthropods per unit biomass or per unit area is not an adequate measure of a system's resilience. What matters is the trophic relationships among the arthropods and other organisms.

Zadoks: The entomologists teach me that the number of predators and parasitoids is 10–100-fold the number of plant-damaging insects, so numbers do matter.

Jayaraj: In Tamil Nadu, we have been conducting IPM demonstrations in farmers' fields in many districts involving four or five major crops for the last decade. Our observations indicate that for one pest at a given stage (egg, larva or adult), there are 8–12 natural enemies in a typical area managed by IPM as against one or two natural enemies per pest in a non-IPM village. So the farmers are aware of the usefulness of IPM and slowly this practice is spreading.

Unfortunately, there are several other systems which interact with this— suddenly, the balance of life is disturbed and one pest dominates. The farmers then resort to pesticides. They may choose the wrong pesticide and it is often applied excessively owing to wrong advice from intermediaries, including pesticide dealers, over whom we don't have much control in India. This leads to the catastrophic situation.

So IPM should be adopted much more rapidly, but this is not occurring, because of socioeconomic problems. Farmers and villages are not able to adopt IPM consistently. Social scientists will have to study this more comprehensively rather than looking at only a few components of IPM. If you listen to the radio or television, read newspapers or hear through personal contacts, very often it is said that we have reached the stage of need-based chemical control, according to an economic threshold. The message stops there; there is no mention of the different components of IPM that could be applied. Only under supervisory programmes, involving entomologists and plant pathologists, is IPM fairly successful. We need to discuss further how to implement the IPM concept for sustainable agriculture.

Bhattacharyya: The definition of sustainability differs from region to region. A common definition would be: internal solution of the internal problem. This definition will solve many of the ambiguities and narrow the difference between countries with different interests. When we talk about an internal solution of the internal problem, we see the importance of involving people. It is the people who know the internal system. Researchers, social scientists, including agricultural scientists, in a particular system are still outsiders. If you really want to know a system, it is better to involve the people. That includes even the gender issue of whether the men or women should be confronted with a particular problem. We should also keep in mind that in many Third World countries the demands for inputs are created demands.

Rabbinge: The positive definition of sustainability is always very difficult. I think it's better to say what is unsustainable. Things such as depletion of nutrients, overuse of pesticides or overuse of nitrogen are quite clear and we know they should be avoided. Sustainability is very dependent on the way people, institutions or governments perceive risk. Risk, not only in ecological terms but also in social or economic terms, has a considerable influence on the way we would like to deal with different systems—either at the level of the cropping system or at higher integration levels. Many things are uncertain: the species diversity, how ecosystems operate, what you would like to maintain, how you would like to maintain it, and at the cost of what. Sustainability as such should be made much more operational before it can be discussed in detail. For the time being, I think we should concentrate on eliminating unsustainability.

Swaminathan: The definition of sustainability is less important to me: we all agree it's a complex factor, it's a location-specific factor. The main thing is how to ensure advances in productivity without undue harm to the environmental capital stocks. When I joined IRRI in 1982, I looked at the yield data for 1966–1981. At the beginning of that period, certain fields were yielding about $19 \, \text{t ha}^{-1}$ of rice per year. Over 20 years, the annual drop in maximum yield was $100 \, \text{kg ha}^{-1}$. We initiated a multidisciplinary experiment to ask why, in spite of all the scientific expertise, maximum yields were falling. Studies over six years showed that a major reason was a gradual reduction in the durations of the varieties. Breeders were concentrating on productivity per day rather than per crop. In addition, pest problems, particularly tungro virus, were getting serious.

Following that, we had a consultation on the definition of sustainability. The group consisted of economists, social scientists and technologists. They all agreed that measurement of sustainability requires the integration of at least two principle parameters. One is the economic parameter, which drives farmers' investment decisions. As Vineeta Hoon said, farmers know certain things are unsustainable, but they may have no option, because they have to make a living. Awareness of alternative options really doesn't help; farmers must have the economic capability to choose among options. We agreed one good way of measuring economic sustainability is output/input costs. If you spend $100 and get $1000, then you have an index. This has to be multiplied by a suitable factor to get an index of the impact of the technology on environmental capital stocks.

This is where the difficulty came in: how do you measure this impact? Dr Robert Repetto at the World Resources Institute, Washington, examined the national income statistics of Indonesia, using methodologies for measurement of the gross domestic product and net domestic product (Repetto et al 1989). The net domestic product, which takes into consideration the depletion of natural resources such as forests, was much lower than gross domestic product. Several such methodologies are now emerging. Our centre is called the Centre for Research on Sustainable Agricultural and Rural Development. How do we measure the impact of our work in the villages? There must be indicators by which we can say whether we are progressing towards sustainability or unsustainability. When I was President of the International Union for the Conservation of Nature and National Resources (1984–1990), we struggled hard to develop a definition of sustainable development. Finally, a book called *Caring for the Earth* (IUCN et al 1991) defined sustainable development as improving the quality of human life while living within the carrying capacity of the supporting ecosystem. We concluded that sustainable lifestyle is fundamental to sustainable development. This has a powerful message for industrialized countries that unsustainable lifestyles can only erode natural resources. On the other hand, for the one thousand million people living in poverty and struggling to have any life at all, the need to improve the quality of human life is extremely urgent.

In our centre we have developed a sustainable livelihood security index, which has three dimensions (M.S. Swaminathan Research Foundation, Annual Report 1991–1992, Madras, India). The ecological dimension is measured by the carrying capacity at the current level of technology. People themselves can measure this. If the level of technology is improved, for example if unirrigated land becomes irrigated, the carrying capacity is enhanced. The second aspect is the livelihood of the family considering multiple income sources, i.e. sources of on-farm and off-farm income. In countries like India and China, the only way to help people realize their own wish to maintain natural resources in a sustainable way is to provide multiple sources of income to the family. An average family may have five or six members. Families that have some non-agricultural income have greater livelihood security and they are more careful in the exploitation of natural resources. The industrialized countries have withdrawn most of their people from farming, but in the developing countries most people depend on farming enterprises. The third dimension is female literacy. We find that wherever women have both education and an opportunity for skilled employment, the natural resource base is safe guarded. Where women are concerned, over-exploitation of natural resources is due to economic and livelihood pressures rather than greed or indifference. The poorer the population, the greater is the dependence on common property resources for survival.

So one can go on searching for new definitions, but what is important is to develop methods of measurement, which relate to the particular social, ecological and economic conditions prevailing in a given area. In our measurement methodology, we have tried to integrate considerations of ecological sustainability with those of economic viability and social equity.

Varma: The off-farm income is very important in countries like India with an increasing population. In India, nearly 70% of the people still live in the villages and farming supplies their major income. This is about 580 million people. The rising population has also resulted in a reduction in the size of farms (Varma 1989). The problem is how to sustain agricultural production on small farms to meet the rising demand.

Zadoks: The gender issue is relevant, even in a country like The Netherlands. In developing countries this is far more important, because women form the unseen labour force.

Escalada: The Rice IPM Network, with support from the Swiss Development Cooperation, is assisting the Department of Agricultural Extension in Thailand to address the changing role of women in rice pest management. The situation in central Thailand is unique because there's a very strong rural migration, which has forced women farmers to take on major tasks in rice farming, such as pest management, which was outside their previous domain. The men who have been trained in pest management have to work overseas or have moved to urban centres for employment.

To assess the key issues affecting women's participation in rice pest management, the Rice IPM Network supported diagnostic interviews and a formal survey in central Thailand. Subsequently, a workshop on women's role in rice pest management and training curriculum development was organized. These interviews and survey revealed that women generally have poor knowledge and lack basic skills in the diagnosis of pests and diseases, choice of pesticides, methods of application and concepts of natural control. They are also unaware of the chronic effects of pesticides.

Because pest management is a recent responsibility of women, it presents an opportunity to teach them the right concepts at the formative stage. While the training requirements of men and women are similar, several aspects may need special attention. Training is regarded as time consuming and most women are not willing to participate because of their commitments to household duties. Season-long training in IPM would be a burden to them. Since training has to fit into women's busy schedules, the use of farmer participatory research to improve perceptions of pest management and practice might be appropriate.

Zadoks: It might be interesting in this respect to look at the situation in southern black Africa, where the majority of farmers are women.

Neuenschwander: There was a good slogan on one of our pamphlets—meet the African farmer and her husband!

Othman: In Malaysia, we are trying to teach women how to be good leaders, how to raise the standard of living in the farm families and how to be entrepreneurs. At my training centre, they are developing modules with the university people. There are master trainers who go back and train the farmers' wives, in how to be a good leader, how to help their families and increase their standard of living, and how to be an entrepreneur.

Kenmore: In Chingleput district in Tamil Nadu, there's been a fairly long-term project on women in development supported by DANIDA. This worked with groups of women in villages by focusing on women already working as staff in the extension system. The Indian national IPM system that Dr Ragunathan directs found that female extension workers were extremely enthusiastic about new ideas. They were not daunted by complexity; they were mostly interested in getting hold of information from which they had previously been structurally barred. It took an intervention programme to give them that opportunity.

References

IUCN, UNEP, WWF 1991 Caring for the earth: a strategy for sustainable living. Earthscan, London

Repetto R, Magrath WB, Wells M, Beer C, Rossini F 1989 Wasting assets: natural resources in the national income accounts. World Resources Institute, Washington, DC

Varma A 1989 Manpower development and utilization in agricultural research. Presidential address. In: Programme of agricultural sciences section, Indian science congress, Madurai, 1989. p 1–25

Wade R 1988 Village republics. Oxford University Press, Oxford

Crop protection: why and how

J. C. Zadoks

Department of Phytopathology, Wageningen Agricultural University, PO Box 8025, NL-6700 EE Wageningen, The Netherlands

Abstract. The answer to the question 'why' has proximate and ultimate roots. The proximate answer, 'pests take our harvest', compels one to act. Crops and their pests are products of domestication, the ancestors of the pests still exist in nature. Eradication is impossible and undesirable. What action should be taken? The ultimate answer, how organisms became pests, may tell us how to act. Old cropping systems had man-made ecological sustainability but do not have economic sustainability in modern times. To achieve both we must prevent rather than control pest outbreaks, using old and new ecological tricks, with application of pesticides in emergency cases only. The action will require a move from chemistry to ecology. Optimism on regaining some ecological sustainability is mixed with doubts on economic sustainability. Can farmers be asked to invest in the future at the expense of today's family income? Crop protection faces new technical and moral problems.

1993 Crop protection and sustainable agriculture. Wiley, Chichester (Ciba Foundation Symposium 177) p 48–60

Crop protection: why?

The answer to this question has proximate and ultimate roots. The proximate answer is that 'pests take our harvest'. This answer is simple and compels us to act. The ultimate answer is that 'pests were domesticated together with their host plants'. This answer is complex but it may tell us how to act.

We tend to believe that old cropping systems had ecological sustainability, albeit a man-made sustainability. Evidence is found in classical rice cropping systems (Padilla 1991), now nearly derelict. In olden times, these systems also had economic sustainability, but not nowadays.

At all times, agriculture responds to the needs of society. So do cropping systems, crops and crop pests. But whereas cropping systems and crops usually respond in a positive sense, the response of pests may be negative, at least in the eye of the grower.

Developments in agriculture

We will take a rather distant look at our subject matter. The word 'pest' will be used in a general sense, including birds and rodents, insects and mites, nematodes, fungi and bacteria, viruses and viroids. The ancestors of the pests

48

still exist in nature. Eradication is impossible and undesirable. All so-called pest organisms have their own natural place in the world's ecosystems and any organism may develop into a pest.

Table 1 mentions some developments relevant to crop protection. Analysis of these developments may suggest answers to the question 'Crop protection: how?'

Enlargement and aggregation of fields favour the rapid spread of pests and usually hamper natural enemies of these pests. Enlargement and aggregation are ongoing processes governed by hunger for land and the needs of mechanization, but in a secular perspective ebb and flow alternate. In the European Common Market, shrink and disaggregation are no longer unmentionable (Netherlands Scientific Council for Government Policy 1992).

Genetic uniformity of crops may be considered at three levels—the species, the cultivar and the genotype. The Western World has moved towards extremes of uniformity within and between fields, with equally extreme vulnerability to pests (National Academy of Sciences 1972). The Third World has stuck to mixed cropping: its beneficial effects on crop protection are being unravelled and classical wisdom is being followed once again. The suppressive effects of cultivar mixtures on diseases have been recognized. Unfortunately, multiline varieties are not economically feasible. The typically Western dogma of uniformity may serve the supplier and the customer, but not necessarily the grower. The dogma's general validity must be questioned.

The increase in plant density is due to the combined effect of improvements in soil tillage, water management, plant nutrition (especially high nitrogen levels) and plant breeding. It has led to higher humidity within the canopy and to concomitant changes of pest species. This development is not irreversible. New idiotypes of crop plants, new crop structures (as affected, for example, by distance between rows) and new machinery may be developed.

The increase in harvest index (the harvestable fraction of above-ground biomass) has been the corner-stone of the 'magic rice' and the 'wonder wheat' of the green revolution. This major advance made yields more vulnerable to pests, because crops lost their ability to suppress weeds and plants lost

TABLE 1   Changes in agriculture relevant to crop protection

Enlargement of fields.
Aggregation of fields.
Genetic uniformity of crops (species, cultivar and genotype levels).
Increase of plant density.
Increase of harvest index.
Specialization.
Mechanization.
Exchange of infected material (seed, plants, soil).

Adapted from Zadoks & Schein (1979).

their capacity to compensate for bruises and damage caused by pests. Should we move one step back?

Specialization in one or a few crops has contributed enormously to the labour efficiency of Western agriculture. The example of The Netherlands, where soil-borne pests became an overwhelming problem, contains a warning for other parts of the world. Where traditional specialization in rice showed no obvious problems, recent specialization in vegetables on South-East Asian hill sites seems disastrous (van Keulen et al 1994).

Mechanization dramatically reduced labour needs in agriculture but the use of modern heavy machinery has several disadvantages, of which soil compaction may be the worst. In potato cultivation, mechanization solved some pest problems but caused many new ones. These are now met by 'soft' mechanization, including biological control (Mulder et al 1992).

Finally, the free international exchange of infected or infested plant materials, be it by the seed trade, scientific institutions, individual scientists or tourists, has greatly contributed to agriculture's crop protection problems. Quarantine and containment alleviate some of the problems, at least temporarily.

Each of the items of Table 1 has its ecological implications. For each item, farmers and authorities may have the freedom to decide how far they want to go in one direction or the other. Conventional agriculture and organic farming have chosen opposite directions of development. Their choices have little to do with technology itself, since both conventional and organic farming can be high-tech, though in different ways. Physicochemical technology can be opposed to, or better, placed side by side with ecological technology (ecotechnology). Biotechnology may be included for early identification of pests and improvement of varieties.

Agricultural technology criticized

The intensification of agriculture has gone on for centuries. In the Western World, where agricultural science developed in the 19th century, production per hectare increased slowly but steadily at a rate of about 1% per year. This accelerated after World War II to attain a rate of increase of about 70 kg ha^{-1} per year (de Wit et al 1987). This was the first green revolution. All branches of agricultural science contributed—soil science, water management (irrigation and drainage), plant nutrition, plant breeding, agronomy and crop protection.

In the Western World the economic pressures are such that farms grow bigger, the number of farmers decreases and overproduction increases. High production levels are obtained by means of high energy inputs under whatever disguise—fuel, fertilizer, pesticide or equipment. New technology is rapidly incorporated. The stage has been reached at which technology is seen as the driving force behind the continuing increase in production (de Hoogh 1987).

In contrast, in South and South-East Asia, the driving force was population growth. With a land availability of about 0.25 ha per capita, production had to increase. The second green revolution, introduced by the International Agricultural Research Centres, came just in time. It rapidly overcame its first generation of problems, severe outbreaks of sometimes new pests. The basic philosophy was simple, 'put it all into the seed', hence the emphasis on plant breeding. The genotype should have broad adaptability and high productivity. Given water, fertilizer and pesticide, a rich harvest would flow in automatically.

Unfortunately, life is stronger than doctrine: not all expectations came true. Old and new pests became epidemic, such as tungro in rice. Pesticide-induced pests, such as the brown plant hopper, caused severe damage. For many farmers the inputs are too expensive, as some picketing farmers explained during the 1987 International Plant Protection Congress in Manila. Alternatively, the credit to buy the inputs is too expensive. Farmers get poisoned by pesticides and become aware of their hazards (Loevinsohn 1987). More classical ways of intensive farming, for example combining rice and fish, or rice and ducks, have become impossible because of overusage of pesticides. At present, public concern with environmental issues is growing and will eventually place constraints on farming. In short, the green revolution doctrine is beginning to fade.

Nevertheless, the green revolution with its high input agriculture has served the world well, though temporarily. Pesticide-induced pests, ill health due to pesticide and nitrate intoxication, environmental pollution by pesticides and fertilizers have reached unacceptable levels, threatening not only today's people but also future generations. Hence the idea of 'durable', 'sustainable' or 'integrated' agriculture.

Toward sustainable crop protection

Though the terminology may seem confused, the intentions are clear. The terms 'durable' and 'sustainable' point to the wish that the state of the means of primary production (soil, water and air) of our children and grandchildren will be at least as good as it is now. 'Integrated' has a similar connotation but points towards the widened objectives of modern agriculture. These are not merely production of food, but also production of renewable resources (soil amendments, energy, chemicals), care of the landscape and maintenance of biodiversity. These items are becoming explicit objectives of agricultural policy in The Netherlands and, with some delay, in the rest of Europe.

The question 'Crop protection: how?' is well answered in The Netherlands' Multi-Year Crop Protection Plan (Anonymous 1991). The plan states three policy objectives: (i) reduction of dependence on pesticides; (ii) reduction of volume of pesticides used; and (iii) reduction of emission of pesticides. These policies are now being implemented, in part by developing new (eco)technology.

Modern crop protection science has contributed a major solution, that of integrated pest management (IPM). According to the FAO report of 1968,

> . . . Integrated Control . . . is defined as a pest management system that, in the context of the associated environment and the population dynamics of the pest species, utilizes all suitable techniques in as compatible a manner as possible and maintains the pest populations at levels below those causing economic injury. In its restricted sense, it refers to the management of single pest species on specific crops or in particular places. In a more general sense, it applies to the coordinated management of all pest populations in the agricultural or forest environment. It is not simply the juxtaposition or superimposition of two control techniques . . . but the integration of all suitable management techniques with the natural regulating and limiting elements of the environment.

The underlying concept was in no way new but stating it explicitly has wrought wonders in terms of reorientation of scientific research and governmental policy. In the developing world, spectacular results were obtained in soybean in Brazil and in rice in South and South-East Asia, at incredibly low costs of investment (Zadoks 1993). Farmers profit from IPM and consumer prices will hardly increase.

Table 2 shows when and where governmental policy picked up the message. South and South-East Asia, where political leaders embraced IPM early and explicitly, take the position of honour. In several European countries, sustainability-oriented decisions were taken without mentioning IPM explicitly, (Federal) Germany being the positive exception.

Now that the International Agricultural Research Centres also see that they cannot put everything into the seed, and study and advocate IPM, the message

TABLE 2   Integrated pest management declared as official policy

1985	India	Ministerial declaration	+
1985	Malaysia	Ministerial declaration	+
1986	Germany	Parliamentary decision, Plant Protection Act	+
1986	Indonesia	Presidential decree	+
1986	Philippines	Presidential declaration	−
1987	Denmark	Parliamentary decision	−
1987	Sweden	Parliamentary decision	−
1991	The Netherlands	Cabinet decision: Multi-Year Plan for Crop Protection	−
1992	United Nations Conference on Environment Development	World's Heads of State, Agenda 21, Rio de Janeiro	+

+, explicit statement; −, implicit statement.

will spread rapidly through the Third World, whereas in the Western World environmental pressure will force agriculture to adopt IPM. But what about the former socialist countries? China is a leader in IPM, the former Soviet Union had extensive biological control programmes, and the former German Democratic Republic used mixed cultivars of barley to avoid purchase of pesticides from the West.

IPM is a firm step on the road towards sustainable crop protection, but not necessarily the last step. The Indonesian Rice Programme has developed beyond the stage of IPM and aims at healthy crops. Crop health and ecological equilibrium are the catchphrases. Organic farming is developing strongly in many parts of the world and by producing new ecotechnology will open new avenues towards crop health.

In may be too early to attach labels to these new developments. One was called integrated crop management (ICM), first by Heitefuss (1989):

Integrated crop management embraces systems of plant production most appropriate to the respective location and environment, in which, in consideration of economic and environmental requirements, all suitable procedures of agronomy, plant nutrition and plant protection are employed as harmoniously as possible with each other, utilizing biological-technical progress as well as natural regulatory factors of noxious organisms in order to guarantee long-term assured yields and economic success.

then by the FAO (1991)

Integrated Crop Management (ICM) embraces all activities in the production system and is composed of several management activities focusing on particular constraints, such as integrated pest management (IPM), integrated nutrient management (INM), integrated water management (IWM), etc. ICM is concerned with managing a production system to optimize the use of natural resources, reduce environmental risk and maximize output. The goals of a particular management system are dependent upon natural, socio-economic and technological resources and their inter-relationships.

Though the obvious advantages of good, modern pesticides are not brushed aside, it emphasizes plant health through prevention. This may appeal to farmers who, in order to survive, are experienced agricultural ecologists. A now classic example of prevention is seed certification, which still makes a major contribution to crop health. It is not accidental that a farmer-scientist, the Dutchman Oortwijn Botjes, invented the idea of potato seed certification in 1920. Listen to farmers; they have ideas.

Plant health

We want healthy crops, for healthy people, in a healthy environment. We may need to sacrifice the top fraction of the yield for that purpose. The sacrifice may hurt the pride of the farmer but it will not hurt his or her purse. The mentality of crop protectionists, growers and the public will change (Zadoks 1991). Agricultural science is setting itself the task of realizing the concept of sustainability, but in several instances practice has found its own way to sustainability. Renewed interest in crop husbandry, biological control, IPM and habitat management are all steps on the road, along which ICM may be the next station. Let us make the green revolution greener (Swaminathan 1990).

Ecological sustainability should be coupled with economic sustainability. Optimism on regaining some ecological sustainability is mixed with doubts on economic sustainability. Where change is needed, investment precedes profit. Can farmers be asked to invest in the future at the expense of today's family income? Crop protection faces new technical and ethical issues.

References

Anonymous 1991 Meerjarenplan Gewasbescherming. Regeringsbeslissing (Multi-year crop protection plan. Cabinet decision; in Dutch). Sdu Publishers, The Hague
de Hoogh J 1987 Agricultural policies in industrial countries and their effects on the Third World. A critical view on the comparative-static analysis of a dynamic process. Tijdschr Sociaalwetensch Onderz Landbouw 2:68–81
de Wit CT, Huisman H, Rabbinge R 1987 Agriculture and its environment: are there other ways? Agric Syst 23:211–236
FAO 1968 Report of the first session of the FAO panel of experts on integrated pest control. FAO, Rome (Meet Rep PL/1967/M/7)
FAO 1991 FAO/Netherlands conference on agriculture and the environment, 's-Hertogenbosch, The Netherlands, 15–19 April 1991. Sustainable crop production and protection. Background document 2. FAO, Rome
Heitefuss R 1989 Crop and plant protection. The practical foundations. Ellis Horwood, Chichester
Loevinsohn ME 1987 Insecticide use and increased mortality in rural central Luzon, Philippines. Lancet 1:1359–1362
Mulder A, Turkensteen LJ, Bouman A 1992 Perspectives of green-crop-harvesting to control soil-borne and storage diseases of seed potatoes. Neth J Plant Pathol 98 (suppl 2):103–114
National Academy of Sciences Committee on Genetic Vulnerability of Major Crops 1972 Genetic vulnerability of major crops. National Academy of Sciences, Washington, DC
Netherlands Scientific Council for Government Policy 1992 Ground for choices. Four perspectives for the rural areas in the European Community. Sdu Publishers, The Hague (Rep Gov 42)
Padilla H 1991 The Bontoc rice terraces: high and stable yields. ILEIA News 7:4–6
Swaminathan MS 1990 The green revolution and small-farm agriculture. In: CIMMYT 1990 annual report. CIMMYT, Mexico, p 12–15

van Keulen J, Schönherr IA, Zadoks JC 1994 Key causes of pesticide misuse in highland vegetable production in Southeast Asia. Trop Pest Manage, in press

Zadoks JC 1991 A hundred and more years of plant protection in The Netherlands. Neth J Plant Pathol 97:3–24

Zadoks JC 1993 The costs of change in plant protection. J Plant Prot Trop 9: 151–159

Zadoks JC, Schein RD 1979 Epidemiology and plant disease management. Oxford University Press, New York

DISCUSSION

Jeger: The effect of plant density will be a very important question for areas like agroforestry and biomass production. There seems to be an implicit consensus that increases in plant density *per se* will normally exacerbate plant disease and pest problems. Burdon & Chilvers (1982) reviewed the effects of plant density on plant diseases. Although increasing plant density generally favoured disease development, there were several cases where there were either neutral effects or disease was less apparent at high plant densities.

Zadoks: Plant density can have any effect. On the whole, humidity within the crop is stimulated at higher densities, so foliar diseases may increase, while foliar insects may decrease. However, if the plant density is high, the dispersal of fungal spores becomes so difficult that the fungus can no longer spread.

Waibel: You proposed the term 'sustainable crop protection', which I am sure most of the people in this room like very much. You said crop losses have increased in absolute terms, not in relative terms. This implies that the increase in loss is due to a shift in the attainable yield. This would mean that, because most pest problems in the past were solved using pesticide-based solutions, the productivity of pesticide use has not declined. If it had, losses would have increased in relative terms. So a danger inherent in the use of a term like sustainable crop protection, taking into consideration the political economy in which we live, is that a pesticide-based crop protection system could be sustainable in the sense that it would always create its own market.

Zadoks: I introduced the term sustainable crop protection, and I am happy if you try to shoot it down. There is no term that has no disadvantages. I am not really against pesticides: I am in favour of a good use of pesticides. In the future we will certainly have new pesticides and/or new methods of application, such as seed coating techniques, which will be much better from the sustainability point of view than those we have now.

Wightman: I would like to congratulate Professor Zadoks on his talk. I am relieved and pleased that many people are aware of some of the broader issues, but I would like to mention an opposite case. In Thailand, I understand, consumers are being offered pieces of paper explaining how to make the food and vegetables safe from pesticides. There are so many chemical residues in the fruits and vegetables that this is necessary. That is the perspective being adopted in a developing country.

One of the developments of the plant breeders was the introduction of dwarf plants, which were successfully adopted in Europe. Sorghum farmers in both Africa and India have been offered dwarf varieties with potentially high grain yields and have refused them because they can't use the straw for thatching and so on. The same has occurred with pigeon pea. The breeders have produced dwarf pigeon peas which are rejected by farmers because of their susceptibility to insects and the fact that they can't be used for firewood, for fences and so on. We are kidding ourselves if we think of IPM as a closed box. We have to think in terms of an integrated crop protection system within an integrated farm management system.

Zadoks: The dwarf varieties of rice are also resisted by the farmers in Bangladesh, because they need the rice straw as cattle fodder. I think dwarf varieties were successful in Europe because we no longer needed the straw. The straw used to be utilized for thatching, making pillows, litter for the cattle and it was added to the soil. Materials like cotton and kapok have replaced straw for these purposes. If you look at pictures in medieval books, the people and the crop were equally tall. The people were a little shorter than they are now, but the essence of the message is that the crop was so tall because people needed the straw as much as the grain.

Mehrotra: Dwarf varieties are being accepted in some cases. Indian farmers readily accept dwarf varieties of wheat but for sorghum they prefer tall varieties that provide fodder for cattle.

Although IPM has been an official policy of the government of India since 1985, I am afraid that it is not practised by farmers to a great extent, except on government farms. In a country such as India, where agricultural practices vary from north to south, east to west, and district to district, the resolution of the government to use IPM is very hard to apply universally.

Varma: The fragmentation of fields in India has also made it difficult to apply IPM practices on individual farms. They need to be integrated on a larger basis.

Neuenschwander: IPM is a beautiful concept and, like apple pie or motherhood, everybody is in favour of it and that is its danger—the label often gets misused. IPM has been developed to counter the abuse of insecticides. It promotes the reasonable use of insecticides in conjunction with three other methodologies, namely host plant resistance, cultural practices and biological control. I've been concerned with several projects where specific cultural practices, crop varieties and insecticides were already in use. I was told 'Now put in biological control'. It didn't work. IPM doesn't work if you add biological control to other methodologies piecemeal.

IPM has to be based on biological control first, because this is the component that is usually least manageable. Unfortunately, it is also often the least well known.

By contrast, insecticides come with proper information on their use provided by the salesman. Knowledge of cultural practices or different crop varieties may be with the farmer, but not in the scientific community yet, or vice versa. The research institutes have an important role to play in developing and assimilating that knowledge. At the International Institute of Tropical Agriculture (IITA) we try to provide the ecological and biological basis for IPM and then integrate cultural practices and crop varieties. We place little emphasis on insecticides, because their use is often uneconomical and—as it turns out—not needed.

Mehrotra: In India, we have been using not more than 450 g of technical material per ha per year, which is rather low. Nevertheless, there has been abuse of pesticides. This has created a horrible situation of pesticide resistance in insect pests. We have not used much fungicide and there is no indication of fungicide resistance to date. There are indications, however, that apple scab may develop resistance in the near future because of the extensive use of fungicides to combat it.

Varma: It's true that in India pesticide consumption is much less than in many countries. The problem is that pesticides are used on very selected crops, so there is relatively intensive use in some areas, even though the overall usage is low.

Jayaraj: I would like to comment on crop mixes and cultivar mixes. In India there are frequent outbreaks of pests and diseases where a crop is raised in monoculture with very few genotypes. In parts of peninsular India, rice, sugarcane, banana and other crops are grown simultaneously. Several different varieties of rice are also grown. There we find greater stability against pests and diseases. Mixed cropping should be promoted for sustainable crop protection.

Secondly, there is the question of exchange of infected materials. Though there are quarantine systems in many countries, particularly in the Third World, these are not very effective. A big country like India requires domestic quarantine. Sugarcane scale from Tamil Nadu spreads with the planting material to the whole country in one season. This is a serious problem.

Lastly, even though IPM has not been adopted extensively in India, the awareness that we should minimize the pesticide load was present even at the beginning of the century. Tamil Nadu, formerly called Madras Province, enacted the Madras Insect Pest Act in 1912. This was followed by the Government of India's Insect Pests Act in 1914. These acts listed a number of measures to prevent the spread of pests and diseases and to minimize the losses caused by them. These measures are no longer being implemented: the Pest Act is no longer effective. We should consider the legal issues for promotion of sustainable crop protection.

Zadoks: People have been too eager to promote IPM as a new idea, mainly in response to pesticide resistance. They have not considered that a farmer adopts a new technology (and IPM is just another technology) only if it is profitable. In the greenhouse business, biological control as part of IPM is very profitable. In The Netherlands, we have reconsidered the breeding of greenhouse crops; these are now modelled to improve IPM. So you can interlink biological control and plant breeding.

The penetration of IPM has to come from different sources. The most important source in a westernized country with a completely market-oriented system is consumer demand. Consumer-driven introduction of IPM is proceeding rapidly in north-western Europe. The second source, government-driven IPM, goes very slowly. Farmers have a long-standing tradition of mistrusting the messages of governments. Such messages are usually only mixed blessings at the very best. They are always partly technical and sound and partly political. Farmers have little to do with the official policies. I could give examples where the farmers were right and the extension service was wrong. A third source comes from within the farming community. Farmers themselves are always experimenting, and they see the ill effects of overuse of inputs. They are suffering from pesticide poisoning, especially in Asia. There was one Chinese chap who was very proud that in his district IPM was being implemented. As an argument he said that only one farmer in 10 000 dies every year of pesticide poisoning! There is also evidence that pesticides affect human fertility and farmers are becoming aware of this very fast. Farmers' wives are much more concerned than the farmers themselves about the toxic effects of pesticides. 'Bed sermons', which wives may preach, can be very effective at influencing the husbands.

Norton: You say that adoption of IPM can be consumer driven—how feasible is that in the longer term? I think legislation is the best approach to implementation of IPM. The consumer approach works only if there's a high premium: in other words if farmers are encouraged to go into IPM by getting, say, a 40% premium, just as with organic agriculture. Once more and more farmers adopt IPM, the price premium is reduced and there is no longer an incentive. I wonder what proportion of the population would be prepared to pay 30–40% more for their food, particularly at a time of recession and high unemployment. If you want to reduce pesticide use, the best way is probably by pesticide regulation.

Zadoks: It's the big retail chains who advertise consumer goods produced in an environmentally friendly way. They decide whether or not to buy the products. It is not a matter of price. If the retailers don't approve of the production system, they don't buy the goods.

Royle: The International Organization for Biological Control (IOBC) is developing detailed criteria and guidelines for labelling IPM products. There is a great demand for an IOBC label in some countries, initially in fruit production but extending to arable crop production. This is a difficult process:

definitions and guidelines have to be very explicit. They also have legal implications, which are difficult for a non-professional organization to address. It's early days, but this issue will become more important.

Zadoks: In Europe there are at least 15 different labels to stick on fruits, stating that they are produced in an environmentally friendly manner. The consumer not only wants a safe fruit or product, but also wants to be assured that the production method is safe and environmentally friendly. The degree to which we can trust these labels is still a matter of discussion.

In The Netherlands, this development has gone pretty far for greenhouse production. The growers are requested to do detailed book-keeping on their methods of crop management. The system becomes stricter every year. Each year, one more pest or disease is added to the list of those that should be contained by biological treatment in an environmentally friendly way. The growers have taken the initiative themselves and there is social control. A grower who doesn't follow the instructions is kicked out of the system, and that is a very severe punishment.

Royle: The IOBC is trying to develop internationally acceptable guidelines with a system of policing in individual countries. Many growers want to affiliate to a European-wide labelling system.

I would like to present the results of a survey of over 100 participants at an IOBC conference in September 1991, which identified 10 important factors that hinder the implementation of IPM. The participants were researchers, extension workers and decision-makers—the policy-makers who actually make the decisions about IPM research and its implementation. The issues were:

1) No general policy for integrated production and protection at the European level, the whole regional area.

2) Lack of knowledge of IPM by policy-makers and politicians.

3) Lack of knowledge by farmers.

4) Lack of extension officers well trained in IPM.

5) No or very few economic incentives.

6) Insufficient research funding.

7) Unrealistically high product quality demands and unrealistic zero tolerance (this is very much a European or Western situation of course).

8) High costs of IPM programmes.

9) IPM is not clearly defined for outsiders, so there is confusion about the standards for a labelled product.

10) A general lack of confidence in IPM.

Zadoks: We did a study in Western Europe and reached the same conclusions. Internationally, incentive is lacking: in Europe a common market incentive would be appreciated. I am not convinced that legislation is the answer, although it may be a help. Legislation often has many unexpected side effects, which are disadvantageous. Dr Jayaraj mentioned the spread of pests and diseases after they have been introduced into a country. This is an example of where legislation doesn't work. Apart from quarantine, there are containment strategies. In the US, these may be very strict. The containment strategy against *Striga* spp imported into the eastern US is so strict that it works, but the costs are rather high. It's a common experience that you can introduce legislation and people in the beginning tend to obey the law, but in the end they always break or neglect it. Even the law-maker may neglect the laws because new problems arise all the time.

Reference

Burdon JJ, Chilvers GA 1982 Host density as a factor in plant disease ecology. Annu Rev Phytopathol 20:143–166

The economics of food production

Martin Upton

Department of Agricultural Economics and Management, University of Reading, 4 Earley Gate, Whiteknights Road, PO Box 237, Reading RG6 2AR, UK

Abstract. Although world average food production per person is increasing there are many countries, particularly in sub-Saharan Africa, where production has fallen in recent decades. The economic analysis of the world food problem concerns the dynamics of production, income, growth, demand and trade. The 'law of diminishing returns' suggests that labour incomes fall as population density increases. Capital investment and technological change, particularly with a land-saving bias, can overcome this effect. Such land-saving innovations are less appropriate where population densities are lower, as in much of sub-Saharan Africa. Innovations which reduce risk, such as stress- and disease-resistant crop varieties, may be more attractive to farmers. Communal or government action is required to ensure sustainability of food production; to reduce risk, through price stabilization, possibly crop insurance and contingency plans for famine relief; to promote equity and to ensure competitive market conditions. Public funding of agricultural research is necessary to promote growth in food supplies. If increases in supply do not keep pace with growth in demand, food prices rise, attracting resources into food production. If supply grows faster, food prices and farm incomes fall, driving resources out of agriculture. Resources may not move fast enough to correct imbalances. Primary producers are likely to face deteriorating terms of trade. Linkages between food production and other sectors are weak, so primary exports are not a good basis for economic development. Import substitution strategies may damage agriculture. Structural adjustment regimes have been adopted in some countries to correct imbalances and provide an incentive for farmers to increase production. Associated reductions in public expenditure may have a contrary impact.

1993 Crop protection and sustainable agriculture. Wiley, Chichester (Ciba Foundation Symposium 177) p 61–75

Over the past 30 years the world, on average, has become better fed, despite an increase in population of over 2000 million. According to United Nations' statistics, the world average daily calorie supply *per caput* rose from 2316 kcal in 1961–1963 to 2711 kcal in 1989 (Alexandratos 1988, World Bank 1992). The index of food production *per caput* has risen by 19% over the last 10 years (Fig. 1).

However, these favourable aggregate trends hide the facts that there are 300 to 500 million undernourished people in the developing countries, and that food

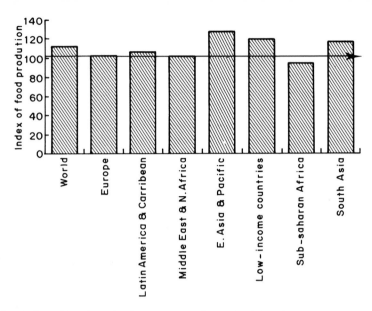

FIG. 1. Index of food production *per caput* in 1988–1990. 100 represents production in 1979–1981.

production *per caput* is falling in some of these countries. In Africa, 30 of the 38 sub-Saharan countries suffered a decline in food production per head over the last 10 years. So too did Asian countries such as Bangladesh, Sri Lanka, Myanmar, Bhutan, Mongolia and the Philippines. Food production *per caput* has also fallen in countries with medium and high incomes in Latin America, the Near East, North Africa and Eastern Europe. There has also been a decline in production in Australia and the United States of America, but this may be intentional, given the problems they face in disposing of agricultural surpluses.

Clearly, there is still cause for concern about the production and distribution of world food supplies. However, economic circumstances differ in different parts of the world and change with time. The economic analysis of the world food problem is seen in terms of the dynamics of production, income growth, demand and trade. This paper therefore deals with three sets of factors: (i) the natural resource base and technological change within the agricultural (food production) sector; (ii) the balance between agriculture and the rest of the economy; (iii) international trade in food.

The natural resource base and technological change

Countries differ substantially in the natural resources available per person, measured crudely as the rural population per square kilometre of agricultural land (crop land and pasture). On this scale, Egypt is the most densely populated, closely followed by Bangladesh, both with over 900 people per square kilometre. Most southern Asian countries and Japan are high on the list. In contrast, much of Africa is relatively sparsely populated with, on average, 39 people per square kilometre (World Bank 1992). Extensive forms of land use, such as pastoralism, ranching or shifting cultivation, are associated with sparse population, low labour inputs and low returns per hectare. Intensive agriculture, in contrast, is associated with high ratios of people to land and high output per hectare (Binswanger & Pingali 1988).

In part, these variations in population density reflect differences in the inherent quality of the land, particularly as this is influenced by the availability of water. In the more arid regions, intensive agriculture may be practicable only with irrigation (see Higgins et al 1982). Nonetheless, the rural population is growing in many parts of the developing world, while the scope for extending the area under cultivation (or grazing) is limited. The 'law of diminishing returns' implies that population growth and increasing intensity of labour inputs on a fixed area of land will result in declining marginal and average production per person. Increasing population pressure also increases the risk that the land will become degraded. Thus population growth may explain the downward trend in food production per person that has been observed in some countries. However, for most countries, population growth has been accompanied by rising food production per person.

The forces for development omitted from the discussion so far are capital investment and technological change. Economists have generally treated these separately. Capital is regarded as a man-made resource; technological change is seen as a flow of new ideas which, when adopted, increase the productivity of land, labour and capital. However, the two are closely linked and their effects are difficult to separate. A change in the quantity of capital used per hectare and per worker generally requires a change in technology. Technological innovations open up new opportunities for productive investment. Indeed, investment is needed in research and development to generate the ideas, in extension communication and education to spread the ideas and in materials and equipment to implement them.

Capital investment and technological change increase the productivity of both land and labour, but may have a bias towards a relative saving in one of these factors. It is generally argued that biological and chemical innovations, such as new high-yielding planting materials or new agrochemicals, increase yields per hectare and have a land-saving bias. Increased labour inputs may be needed to implement such innovations. Mechanization is viewed as having a labour-saving bias (Ruthenberg 1985).

The effect of diminishing returns is that increasing labour input leads to a fall in production per person. Capital investment in new technology, particularly if it has a land-saving bias, enables increased labour input to be associated with

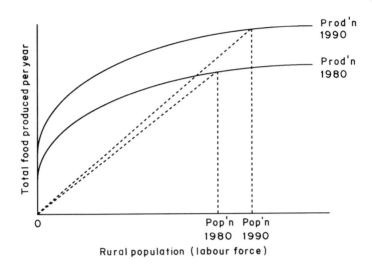

FIG. 2. Diminishing returns and production *per caput*. The effect of diminishing returns is that production *per caput* decreases as the population increases. Improvements in productivity enable production *per caput* to rise, even with a growing population.

higher production per person (Fig. 2). Thus labour productivity (both marginal and average) can increase, even though the agricultural work-force is growing. This has been the achievement of the green revolution in many Asian countries. Such land-saving innovations are less appropriate, and are less likely to be adopted, where population densities are lower, as in much of sub-Saharan Africa. In all countries, investments in land and water conservation may be viewed as land-saving innovations with a long-term benefit in terms of sustainability.

Labour-saving innovations are more appropriate where labour is scarce in relation to the area of agricultural land available. However, in the absence of new areas of accessible virgin land for development, the effect of labour-saving technology must be to displace labour from agriculture. Even in those countries where the population density per square kilometre of agricultural land is low, there may well be substantial unemployment, in which case releasing some of the work-force from agriculture is of no benefit.

Market failure and government intervention

Agricultural production is subject to considerable risk and uncertainty, particularly that due to environmental and climatic variation. The variation, and risk, appear greatest in the more arid areas of the world, under extensive systems of land use, although disasters are possible in more humid regions. In all regions, but particularly the former, food shortages occur sporadically as a result of drought, floods or outbreaks of disease, rather than on a regular annual basis. There is some variation in agricultural production and prices everywhere. Farmers are generally found to be averse to risk. They reduce risk by adopting strategies such as diversification, use of reliable varieties of crops, and production and storage of food for subsistence. Hence, innovations which reduce risk, such as stress- and disease-resistant crop varieties, may be more attractive to farmers than those which simply raise productivity.

The theory of induced innovation suggests that technological change is determined endogenously within the economic system. An increase in population and in the demand for food is reflected in higher agricultural prices. This leads to a rise in the production and supply of food, which in turn raises the demand for, and prices of, scarce resources. The price of the limiting resource rises fastest. Thus farmers are encouraged to adopt technologies with a bias towards saving the limiting resource. It has further been suggested that the allocation of research funds is guided by the relative scarcity of resources, there being an incentive to design technologies which economize on the resource with the fastest rising price. The argument has been extended to incorporate institutional innovations, in land tenure for example (Hayami & Ruttan 1985).

The limitations of this theory are that it concentrates exclusively on the demand for innovation and ignores shifts in the supply of scientific inventions and that it assumes development is driven by market forces. But the market does not stimulate the whole development process successfully and there is a need for government provision of public goods. Four areas merit further consideration: conservation of resources, risk reduction, alleviation of poverty and the provision of rural infrastructure, including agricultural research.

Conservation of resources.   Many aspects of resource conservation and pest and disease control affect areas beyond that in which they are applied. For instance, if farmer A fails to control rainfall run-off on his farm, others further down the slope may experience soil erosion problems, or if he fails to control a pest outbreak, his neighbours will suffer from its spread. Although individual farmers devote resources to the control of erosion or pests where there is a clear private gain from doing so, they are unlikely to take full account of the external effects. Communal or government action is required to provide the public good of the sustainability of the food production system (Fig. 3).

Risk reduction.   Similar arguments apply to risk limitation in agriculture. Although individual farmers adopt risk-avoidance strategies, which have a cost in terms of loss of expected income, the costs can be reduced by risk-pooling and risk-spreading. Institutions such as insurance companies and joint-stock corporations depend upon these principles. However, institutions such as these and markets for risk-sharing are rarely well developed in any economy, least of all in agriculture in a developing country. Therefore, government intervention

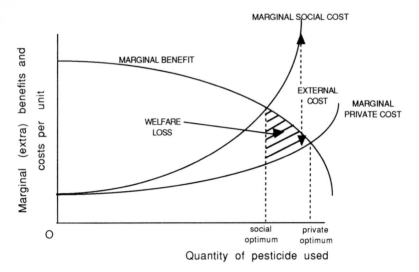

FIG. 3. Welfare loss due to external costs. The use of certain inputs, such as pesticides, has external costs in terms of adverse effects on people other than the user. (For pesticides, these costs might be due to destruction of useful species of animals and plants, development by the pest of genetic resistance to chemical agents, or damage to human health.) If these external costs are not experienced by the user, input usage is likely to exceed the social optimum, thus causing a net loss of social welfare. Government intervention is necessary to limit use of the input and avoid this welfare loss.

is needed to reduce risk through price stabilization, possibly crop insurance, and the establishment of strategic grain reserves or other contingency plans for famine relief (see Gardner et al 1984).

Alleviation of poverty. It has been argued that famines are caused by lack of entitlement or access to food, rather than inadequate supplies (Sen 1981). In short, hunger results from poverty associated with inequality of wealth and income distribution. The operation of free markets, particularly where monopolies arise, may result in a degree of inequality or an incidence of poverty that is regarded as socially unacceptable. Government intervention may be needed to promote a more acceptable distribution of resources and entitlements and to ensure competitive market conditions.

Provision of rural infrastructure. Feeder roads, communications, educational facilities and power and water supplies are characterized as public goods that should be provided communally or publically.

Support for agricultural research and development. The above list may be extended to include the all-important provision of research and development of new technologies for food production. Farmers do experiment and innovate on their own account, but, partly because of the market failures mentioned above and partly because individual farmers and agricultural businesses are unable to exclude others and appropriate all the benefits of research, private investment is inadequate. Public funding of agricultural research is necessary to promote growth in food supplies (see Pardey et al 1989, Wise 1991).

Economic analyses suggest that in many countries suffering inadequate and declining food supplies *per caput*, increased public investment in agricultural research and development would be justified. Yet, publicly funded research has often failed to address the appropriate problems and deliver appropriate technology to farmers. Adoption of promising new technologies has often been hampered by inadequate provision of rural infrastructure and input delivery systems. Adaptive research must therefore be based on a thorough understanding of the farmer's circumstances and conducted through on-farm evaluation, in which both the technical and socio-economic factors are taken into account.

Intersectoral balance

A decline in the proportion of the total population engaged in food production appears to be an inherent feature of economic development. This may involve part-time farming and an increase in off-farm work or migration from rural to urban areas. Nonetheless, for most of the developing world the absolute numbers of people engaged in food production are still increasing. Only in countries with medium and high incomes are the industrial and service sectors large enough to absorb all the annual increase in the national labour force.

The balance between food production and other sectors will be analysed for a closed national economy (i.e. assuming no external trade) in the first instance. From the earlier discussion it may be concluded that the rate of growth in total national food supply, which could be negative in some cases, depends upon the rate of growth of the labour force, investment and technological change in the food sector. It is also assumed that the quantity supplied depends upon the relative price of food. Thus the previously mentioned variables cause continuous shifts in the supply schedule (Fig. 4).

Similarly the factors likely to cause shifts in the demand schedule, over time, are growth in population and income *per capita*. Demand for food grows more slowly than do incomes, as enshrined in Engel's Law. Thus, for a stable population, demand for food grows relatively slowly. At the same time, expenditure on food processing and distribution rises as a proportion of total expenditure on food. However, where the population is growing, so too is the demand for food. If, as a result of these trends, demand for food increases faster than supply, food shortages are likely to result and food prices are likely to rise relative to those for other goods and services. These effects have been observed in recent years in some countries of sub-Saharan Africa. The rise in

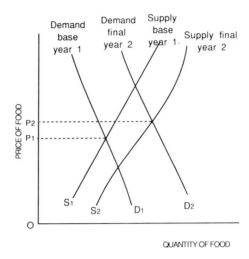

FIG. 4. The relationship between the price of food and production. When demand grows faster than supply, food prices rise.

food prices should provide an incentive for more investment in food production and a decline in migration from rural to urban areas (also see Todaro 1983). Thus, there are forces tending to correct the food deficits. However, as discussed above, constraints on natural resources, market failure and inadequate public funding of research and rural infrastructure may limit the rate of adjustment.

In the apparently more favourable situation, where supply is growing faster than demand, problems of over-supply and embarrassing food surpluses may arise. This has been the situation, created by price support policies, in the European Community. Had farm prices been free to find their own level, they would undoubtedly have fallen and so would have farm incomes. This would have provided incentives for reducing agricultural investment and for farmers and workers to leave the industry even faster than they have done. However, farmers are not motivated solely by financial gain and resist migration to other industries. Much capital investment in agriculture is irreversible. Governments also have objectives other than maintaining an adequate supply of cheap food, such as the aim of conserving the rural environment. For these and other reasons, few agricultural economists believe that reliance on free markets would provide a satisfactory solution to the problems of the agricultural sector.

The discussion here is grossly over-simplified in assuming a closed economy, whereas the European Community is engaged in international trade, as are all nations. It leads to the question of whether the opportunity to trade with other countries influences growth in food production.

International trade in food

Opportunities for trade across national boundaries permit national economies to consume a different range of goods from that produced, which in turn means that specialization in production is feasible. The 'Principle of comparative advantage' (Ricardo 1817, see Heffernan & Sinclair 1990) emphasizes the mutual benefits of trade when each partner specializes in the commodity for which its domestic resource cost per unit of value produced is lowest. It has been argued that low-income economies, lacking capital but with substantial land and labour resources, have a comparative advantage in producing food and other primary products. This suggests that their most appropriate development strategy is growth led by exports of primary products.

However, world demand for food grows only slowly and many of the high-income economies protect their own agricultural industries by limiting imports. Thus primary producers are likely to face deteriorating terms of trade, although the empirical evidence is inconclusive. Another disadvantage of reliance on food exports is the serious instability of world prices and export earnings, which may be exacerbated by the protective policies of high-income economies. Given that linkages between food production and other sectors of most economies are weak, primary exports are not a good basis for the growth and development of the whole economy (see Hirschman 1989).

Many low-income countries have therefore attempted to promote industrialization and to manufacture goods previously imported. This has involved a strategy of industrial protection, through tariffs or import controls and a general bias towards urban areas in policy making. There has also been a tendency for domestic currencies to become over-valued. The effect is not only to discourage exports of primary products, because their value in domestic currency is reduced, but also to encourage imports of food. The adoption of such strategies may be responsible, at least in part, for the relative decline in agricultural production and the debt problem experienced by some countries.

Structural adjustment regimes have been adopted in some of these countries to correct the imbalances. Currency devaluation has raised the domestic price of tradeable goods such as food. This may cause considerable distress and suffering for the landless and urban poor who have to purchase their food but it does provide an incentive for farmers to increase production. Although the supply response for individual commodities is quite strong, in the long run the aggregate response of total agricultural output is limited by constraints on resources. As discussed earlier, technological change offers the best prospects for increasing agricultural productivity. However, this necessitates government funding for research and development and provision of rural infrastructure. Structural adjustment programmes generally involve reductions in public expenditure, but it would be a false economy to cut spending in these crucial areas (Lele 1992).

Summary

Although market forces may induce the adoption of food production technologies appropriate to local conditions and help to achieve a balance between food production and other sectors of a growing economy, market failure may also occur. Government funding is needed for public 'goods' such as research and extension, environmental conservation, stabilization of food prices and supplies, famine relief, and appropriate trade and exchange-rate policies. Enthusiasm for structural reform must not lead governments to curtail too severely their investment in these critical areas. At the same time careful policy analysis is needed to ensure that public interventions are cost effective.

Acknowledgement

I am grateful to Professor Hugh Bunting of Reading University for helpful comments and suggestions on an earlier draft.

References

Alexandratos N (ed) 1988 World agriculture towards 2000. Belhaven Press, London
Binswanger H, Pingali P 1988 Technological priorities for farming in sub-Saharan Africa. World Bank Res Obs 3:81–98
Gardner Bl, Just RE, Kramer RA, Pope RD 1984 Agricultural policy and risk. In: Barry PJ (ed) Risk management in agriculture. Iowa State University Press, Ames, IA, p 231–261
Hayami Y, Ruttan VW 1985 Agricultural development: an international perspective, revised edn. Johns Hopkins University Press, Baltimore, MD
Heffernan S, Sinclair P 1990 Modern international economics. Blackwell, Oxford
Higgins GM, Kassam AH, Marcken L, Fisher G, Shah MM 1982 Potential population supporting capacities of lands in the developing world. FAO, Rome (Tech Rep Proj INT/75/P13)
Hirschman AO 1989 Linkages. In: Eatwell J, Milgate M, Newman P (eds) The new Palgrave: economic development. Macmillan, London, p 210–221
Lele U 1992 Structural adjustment and agriculture: a comparative perspective on performance in Africa, Asia and Latin America. In: Proceedings of the European Association of Agricultural Economists' seminar on food and agricultural policies under structural adjustment, Hohenheim, 21–25 September, 1992. p 3–34
Pardey PG, Kang MS, Elliott H 1989 Structure of public support for national agricultural research systems: a political economy perspective. Agric Econ 3:261–278
Ricardo D 1817 The principles of political economy and taxation. John Murray Publications, London
Ruthenberg H 1985 Innovation policy for small farmers in the tropics: the economics of technical innovations for agricultural development. (ed: HE Janhke) Oxford University Press, Oxford
Sen AK 1981 Poverty and famines: an essay on entitlement and deprivation. Clarendon Press, Oxford
Todaro M (ed) 1983 The struggle for economic development: readings in problems and policies. Longman, Essex, UK

Wise TE (ed) 1991 Agriculture and food research—who benefits? Centre for Agricultural
    Strategy, University of Reading, Reading (Centre Agric Strategy Pap 23)
World Bank 1992 World development report 1992: development and the environment.
    Oxford University Press, Oxford

DISCUSSION

Mehrotra: I totally agree with you about the cost benefits of using a technology. How do you express the external costs of these pesticides, especially the environmental costs, in monetary terms? How many rupees is the use of DDT costing India?

Upton: It is extremely difficult! There are various ways in which a government may intervene to try to limit pollution or harmful effects. The government could charge a tax for polluting; it could pay a subsidy to encourage farmers not to pollute; or the government could lay down quantitative controls. This last approach is the most common, probably because it's the easiest to effect. As you said: how do you estimate the costs? How do you estimate at what level to set a tax or a subsidy? It is difficult.

Mehrotra: The problem is how to convince the bureaucrats. The control of malaria vectors in India is done totally under government supervision. Only three insecticides, DDT, HCH and malathion, are used for this purpose. At present, the two main vectors of malaria, *Anopheles stephensi* and *An. culicifacies*, account for nearly 80% of cases of malaria in India. These two species are now resistant to the insecticides (Sharma 1990). There is a loss of 25–30 million labour days per year because of malaria (Mehrotra 1985). How does one calculate the cost of government action?

Kenmore: A rough estimate of the immediate costs of agricultural pesticide use in India is the loss of over 40 million labour days per year. This is loss due to acute incidents of poisoning from occupational agricultural pesticide use.

Rabbinge: Professor Upton, you mentioned the law of diminishing returns. What are the reasons for this? If you consider the use of external inputs and the increase of productivity from the production ecological side, then you see that when the different inputs are combined in the proper way, they act synergistically. Thus, the law of the diminishing returns applies as soon as you consider only one external input and presume that all the others are not changed. But that's not proper agriculture. Agriculture always involves the combination of inputs—nitrogen, potassium, phosphorus, pesticides, etc. When these are used in the proper way, you have at least the same efficiency and in most cases a higher efficiency, even when working at a higher production level. This is one of the major reasons for the rise in productivity being observed in many countries around the world, independently of their socio-economic structure.

It doesn't mean that the extremes which we have in north-western Europe—the overuse of pesticides, the overuse of fertilizers—should be continued.

The overuse of fertilizers is due to their low cost in comparison with the price of the final product. This is a typical example of market failure. I agree there are different ways of approaching this: governments can try to regulate things completely, which is very difficult. Governments can act through changing the transaction mechanism. Finally, they can control the supply of information, the social infrastructure, and especially the knowledge innovation system in a particular way.

In my opinion, integrated pest management (IPM) is most successful when this knowledge innovation system, including research and extension education, is developed in a proper way. This can then be facilitated by, for example, imposing levies on pesticides.

Upton: On the law of diminishing returns, I was obviously simplifying the issue. I agree that if one could vary all the inputs together, one could overcome the effects of this law. However, I was talking about a situation of an increasing population on a given area of land. In most countries, the area of surplus uncultivated land that's suitable for cultivation is diminishing. So the growing human population has to be supported on a given area of land through increasing production per hectare. In that situation, not all the inputs can be increased together, because the area of land can't be increased.

Rabbinge: It's a question of farming or not farming. The land should be farmed properly and you should try to improve the productivity, taking into account many of the constraints we just mentioned, such that you make the most efficient use of each external input, individually and altogether. I tried to explain earlier the mini-max principle (Rabbinge, this volume): according to this, you try to minimize each of the external inputs, such that you maximize the efficiency of all the other external inputs. Normally, if you are improving the land reclamation level through structural changes, you can achieve a high production level—not the maximum, but fairly high.

Upton: Is it not simply a question of whether there is a limit on total production per hectare or not?

Rabbinge: There's a well defined maximum.

Upton: Well, if there is a well defined maximum, then surely my argument applies? As you increase labour inputs and other inputs per hectare, you approach that limit and there will be a flattening off of the response curve.

Waibel: I am interested in this production function of Dr Rabbinge's. You say that by using the different factors in the right proportion, adding more variables to the production function, you can achieve maximum production. But where in that system are the natural resources? For example, where are the soil microorganisms that deal with nutrients in the soil? Where is susceptibility of the pest to the pesticide? Are they considered in your equation? If not, there will definitely be diminishing returns.

Rabbinge: First, it's not one production function with a lot of variables in it. You have to take into account that cultivating a crop is trying to respond in time, in a dynamic way, to what the crop needs. You have to supply the different requirements at a particular time in a particular ratio. If you do that correctly, you have optimal production. The actual production level achieved is different in different production situations.

The production situation is characterized by the quality of the soil and the level of land reclamation. In general terms, at a low level of land reclamation with poor soils, the optimum of all the external inputs is at a rather low production level. At a high level of land reclamation with good soils, the optimum in terms of efficiency of each of the external inputs is at a high production level. This includes all the self-regulatory mechanisms that you have to take into account, for mineralization, immobilization, self reliance in terms of biological control and things like these.

This is important, because it says that depending on the level of land reclamation, the level of production at which there is maximum efficiency of use of each external input is different.

Nagarajan: For market intervention, I would like to give the example of the subsidy that the government of India provides for pesticides. IPM has now been put on an equal footing. IPM inputs, such as pheromone traps and light traps, have been made accessible to the farmer through subsidies. In this case, we find that IPM, at least in the Coimbatore district, is very competitive. I surveyed the district last month and, to my great satisfaction, I found that IPM works, at least for cotton.

Jayaraj: Whenever you use local inputs for IPM, it generates both employment and income. It is also economically viable, although we have only limited data for limited locations. There is a positive trend among the farmers and entrepreneurs in Tamil Nadu to set up biopesticide units based on our technology. Quite a few unemployed farm graduates as well as some progressive farmers have established production centres for biological control agents. They supply the agents to sugarcane farmers and cotton farmers, using local labour. Similarly, neem seeds are now collected more and more. In India there are roughly 15 million neem trees. So far, we collect only about 20–25% of the seeds. The rate of collection is increasing because of a growing awareness among the farmers of the use of neem-based biopesticides, thanks to the pioneering work done by Dr Ramesh Saxena at IRRI and many scientists all over the world.

We have to promote this concept of local resources for biological control, botanical pest control and so on. Mechanical control is also cheaper in certain situations. Dr Mehrotra has quantified the build-up of resistance to many insecticides in *Helicoverpa* in India. We organized campaigns in Andhra Pradesh and Tamil Nadu using schoolchildren outside class hours to collect and destroy the larvae. This was an economical and safe method. Similarly, rat control campaigns are organized in villages. So wherever labour is available, IPM should include mechanical methods. Otherwise, there is always migration of labour from rural to urban areas, which creates sociological problems.

Zadoks: I'm of the age that I did hand picking of Colorado beetle larvae—10 Dutch cents per jamjar.

Upton: This scheme in Coimbatore sounds splendid, but does it operate entirely in the private sector? Some support is needed in education, in organizing the schemes—either by government or by NGOs, I would suspect.

Jayaraj: The Coimbatore scheme is being organized largely through the NGOs, with support from Tamil Nadu Agricultural University and the Department of Agriculture.

Nagarajan: One aspect that Professor Upton overlooked is intergovernmental intervention. This is equally as important as governmental intervention, so far as developing countries are concerned, because what is of interest to one government may not necessarily be of interest to another government. What may be an environmental issue to one, may not be an issue to another. Therefore I would like to add intergovernmental intervention to your list of intervention under the title of government interaction.

Jayaraj: I would like to request Professor Upton to consider two more areas for the governmental agencies to address. One is quality control of all the inputs supplied, including pesticides and plant protection machinery. The second is the pricing of the inputs. Farmers in Asian and African countries, in particular, cannot afford expensive plant protection. Even if better chemical pesticides were available, they could not buy them because of the high cost. In regions where IPM is working very well, part of the success is due to the fact that pesticides are expensive and beyond the reach of the farmers, except the organochlorines which are available but not that effective. So governments may have to consider pricing as a policy, not only by providing price incentives for food grains but also by controlling the price of the inputs.

Kenmore: From my experience in the field, I should say there's a limit to how far one can adapt any government action or policy. We have found over the last dozen years or so in Asia that it's extremely important to work also with NGOs because of their location specificity. They know the local conditions. They can interpret better, adapt better and innovate better on the local scale, given a government policy signal. It's important that you add these implementing mechanisms to your consideration of the role of government and of the public sector. These mechanisms will not come through the market, I agree completely on that. They have to come through NGOs and small-scale public interest groups that are thriving in Asia.

Upton: The issue of governmental pricing policies comes back to my point regarding possible subsidies or taxes. Taxes on pesticides may be used to restrict their use, or subsidies may be applied to IPM methods to encourage adoption of these. We should be aware of where money flows are occurring, whether subsidies are being applied or whether taxes are being imposed from outside.

Varma: You mentioned land-saving and labour-saving devices, and classified some steps for land-saving, including application of weedicides and new high-yielding crop varieties. I wonder whether we really can classify land-saving and labour-saving methods separately, because they overlap. You require good crop varieties for both land saving and labour saving.

References

Mehrotra KN 1985 Use of DDT and its environmental effects in India. Proc Indian Natl Sci Acad Part B Biol Sci 51:162–184
Rabbinge R 1993 The ecological background of food production. In: Crop protection and sustainable agriculture. Wiley, Chichester (Ciba Found Symp 177) p 2–29
Sharma VP 1990 Malaria: trends and approaches for its control. Presidential address. In: Biological sciences, diamond jubilee session, National Academy of Sciences India, Allahabad, 1990. National Academy of Sciences India, Allahabad, p 1–24

Government intervention in crop protection in developing countries

Hermann Waibel

Institut für Agrarökonomie, Georg-August-Universität, Platz der Göttinger Sieben 5, D-3400 Göttingen, Germany

Abstract. Government interventions in the pesticide market are necessary, particularly in developing countries because of a high likelihood of market failure. The green revolution in Asia made governments introduce programmes which relied heavily on pesticide-based solutions to pest problems. With growing environmental concerns in donor and developing countries, a higher priority is given to natural factors of control. From a review of various types of government interventions it is concluded that government investments related to pest management should focus on reducing the dependence on pesticides rather than concentrating on the minimization of their side-effects.

1993 Crop protection and sustainable agriculture. Wiley, Chichester (Ciba Foundation Symposium 177) p 76–93

The use of chemicals in world agricultural production as a means to ensure high crop yields for a secure food supply requires government involvement in the regulation of the manufacture, trade and use of pesticides. The perceived role of pesticides as an insurance against losses in food crops (which often have serious political consequences) gives them a special role in decision making. Governments in developing countries are caught between 'a rock and a hard place': on the one hand they want to make sure that farmers use enough chemicals, which governments believe will ensure the desired degree of self-sufficiency in food. At the same time, authorities are trying to make farmers use these chemicals judiciously and only when necessary.

The need for government involvement in the pesticide market arises from a high probability of market failure, i.e. market forces do not accurately reflect the true costs of pesticides. Some of the classic reasons for market failure, like externalities and lack of information, are particularly valid for pesticides. Their potential and actual impact on natural resources, such as plant and insect species, air and water, as well as on human health, lead to social costs which are not reflected in the market price of pesticides. Hence if the initiative is left solely to the private sector, the protection of consumers from foods contaminated by pesticides, the protection of the people who apply the pesticides, of domestic animals and wildlife, and of the natural environment in general, cannot be assured. This is particularly true for developing countries where environmental concerns do not always have a high priority. One special problem with pesticides is the difficulty of measuring and demonstrating the full extent of the external costs associated with their use. Some of the negative effects, particularly those on human health, can be assessed only after a very long time. This gives the pesticide problem a special intergenerational dimension, as damage is passed on to future generations.

Historical perspective

Worldwide, in 1988 about US$ 20 thousand million were spent on pesticides, of which 15% was used in Asia not including Japan. If Japan is included, roughly one quarter of the world pesticide use was in Asia. On a crop basis, rice was the most important crop in terms of market share, with 14% of all pesticides used on rice (Anonymous 1990).

Pesticide use in developing countries is coupled with the spread of the green revolution. Before this, farmers relied on their own knowledge and understanding of the ecosystem and made decisions relating to farm practice independently of government influence (Fig. 1). The introduction of high-yielding varieties as a cornerstone of the green revolution was facilitated by national crop intensification programmes. These relied heavily on input packages promoted through credit programmes, which included pesticides as a main component. Recommendations were based on so-called calendar applications and farmers were urged to apply pesticides routinely.

Through credit programmes as a main tool, government interference in on-farm decision making increased. At this stage, governments were trying to tell farmers what was good for them. Extension concepts like the 'training and visit' approach turned farmers into 'consumers of technology'. This approach turned out to be inflexible and was also sometimes ignorant of the rules which govern the ecosystem. Pest outbreaks occurred despite high inputs of chemicals. For example, the brown plant hopper became a threat to the rice crop in several Asian countries. The response of governments was to interfere even more in on-farm decisison making. Government-operated pest surveillance systems were established and additional chemically based interventions were introduced. Practically, governments took control of decision making by enforcing large-scale pest eradication campaigns. The campaigns resulted in high costs, which demanded heavy budgetary outlays. Economic analyses of surveillance programmes in the Philippines and in Thailand showed that because of the high probability of making decisions on the basis of fear of pest damage, resulting in unnecessary pesticide use, the costs of such programmes are likely to exceed their benefits (Waibel 1984, Tuttinghoff 1991). Environmental damage caused by these programmes also became a matter of concern.

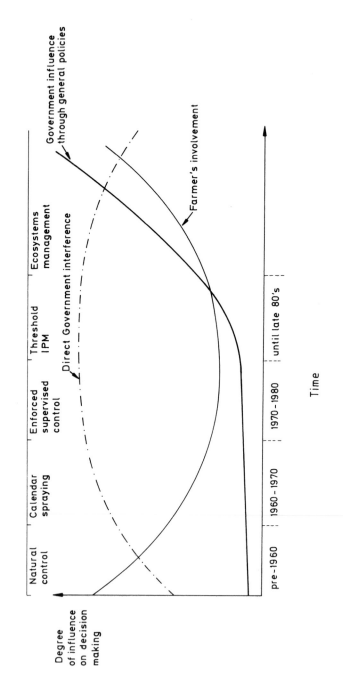

FIG. 1. The role of farmers and governments in making decisions related to pest management in Asia.

The negative effects of over-reliance on pesticides were recognized and the concept of integrated pest management (IPM) was introduced (Brader 1979). Initially, IPM was implemented in a rather top-down fashion. Researchers studied economic thresholds for key pests. The economic threshold is the pest population at which the value of the crop loss prevented by a chemical control measure exceeds the costs of that measure. These thresholds were usually expressed in terms of numbers of pests for scientifically defined sampling units, such as the number of adults and nymphs per plant or the number of egg masses per square meter. Subsequently, government extension workers were given the task of making farmers follow these rigid numbers. By trying to enforce a 'threshold IPM', governments ignored farmers' knowledge of their local ecosystem.

Recently, the success of the Intercountry IPM programme in Asia (Kenmore 1990) has begun to change the role of governments in pest control. In the Indonesian programme, farmers are trained according to the concept of management of the agroecosystem, rather than simply being told to use the technology package of the green revolution. The new approach allows farmers to participate more in on-farm decison making and enables them once again to become the managers of the ecosystems. This restores the situation of the farmers to that which existed before the green revolution, but they now have a greater knowledge and understanding.

The role of governments in decison making is thus shifting away from on-farm interference to more policy-oriented decisons. One example occurred in Indonesia, where 56 pesticide compounds were banned. In general, government interference in decisions relating to pest management is now increasingly concerned not only with the function and performance of the pesticide market but also with improving the entire decison-making process.

Before discussing proposed policy changes, I shall present an overview of the policy instruments available for national government institutions and international organizations and groups.

Policy instruments available at the international level

Roughly 40% of all pesticides produced are traded internationally. 70% of the sales are made by only 10 multinational companies (Knirsch 1990). About 20% of the trade takes place with developing countries, which are mostly importers of these products.

Trade in pesticides differs from trade in other agricultural inputs or commodities. Firstly, although pesticides are classified as agricultural inputs, they are in fact poisons that require special precautionary measures in storage, handling and transportation. This results in additional costs or additional risk, depending on the availability of appropriate facilities, knowledge and skills. Because these vary widely between industrialized and developing countries, regulations and international agreements are necessary.

Global harmonization of pesticide registration is almost impossible because of the variations in national and regional requirements owing to differences in target pests on the one hand and in the relative scarcity of environmental and economic goods on the other hand.

There are two types of intervention with regard to the flow of pesticides from industrialized to developing countries. One is international agreements about trade; the second is guidelines issued by development institutions prescribing the procedure for including pesticides in foreign assistance programmes.

International agreements

International agreements, such as the FAO Code of Conduct on the distribution and use of pesticides, are important policy instruments which counter some of the existing market imperfections through the establishment of rules. The problem with these rules is that their practical implementation depends on facilities, institutions and awareness in the importing countries. To support the Code, the FAO introduced an amendment requiring prior informed consent. According to this, certain hazardous pesticides may not be exported unless the government of the importing country has received full information concerning the regulatory status of that pesticide in the country of export. A key feature of this prior informed consent is a list of pesticides which have been banned or severely restricted for health or environmental reasons in a minimum of five countries. A survey by the Environment Liaison Center International in importing countries, however, showed that although there is general acceptance and support for the principle of prior informed consent, regulatory and enforcement capacities in developing countries are often unable to meet the task with which they are confronted. For example, after being informed about the status of the pesticide in the exporting country, the importing country is given 90 days to react (Pesticide Action Network International 1989). If no reply is received, acceptance is assumed. But it is unrealistic to expect these countries to take all the responsibility for keeping out unwanted imports of hazardous pesticides. Prior informed consent, therefore, can be considered only as an intermediate instrument until the full requirements of the Code of Conduct can be met (Vaagt & Kern 1992).

The type of compounds being traded with developing countries also causes concern. Frequently, pesticides of WHO class Ia and Ib (extremely or very hazardous) are imported by developing countries because they cost less (Rola & Pingali 1992). For example, 60% of all pesticides imported by Thailand belong to these two groups of compounds (H. Waibel, unpublished paper, Workshop Environ Health Impact Pestic Use Rice Cult, Los Banos, 28–30 March 1990).

Agency guidelines

Donor agencies such as the World Bank and the Asian Development Bank, and bilateral agencies, such as US-Aid and the German Agency for Technical Cooperation, have their own policy guidelines regarding the release of pesticide supplies as part of their development programmes. Similarly, these agencies have policies with regard to the planning and implementation of plant protection programmes in developing countries.

Existing guidelines of development banks (World Bank 1985, Asian Development Bank 1987) relate to the selection and procurement of pesticides in projects financed by them. Similar guidelines are being prepared by the Organization for Economic Cooperation and Development (OECD, unpublished paper, Draft working party on development assistance and environment. Good practices for pest and pesticide management, Paris, 1992). Such guidelines usually do not address the level of pesticide use but rather emphasize the way in which pesticides are selected. An analysis of the implementation of the World Bank guidelines on pesticides for 1985–1988 (Hansen 1990) revealed that bank loans were made with no regard for the explicit aim stated in the guidelines 'to reduce reliance on chemical pesticides'. In a survey of nine projects, Hansen found that instead of reduced pesticide use the technical packages financed by these projects often result in increased application.

Policy instruments available at the national level

Nationally, agricultural policies relevant to pest management can be grouped according to their main targets (Fig. 2). One group consists of basically non-pesticide policies aimed at the prevention of pest occurrence within the country; the second is concerned with regulating the function and performance of the pesticide market.

Non-pesticide policies

Non-pesticide policies are designed to reduce the probability of attacks by pests and are therefore damage-avoiding measures. They are mainly quarantine services, breeding of resistant species and crop diversification strategies.

Quarantine services. In the proposed OECD guidelines on pest management in developing countries (OECD, unpublished paper, Draft working party on development assistance and environment. Good practices for pest and pesticide management, Paris, 1992), quarantine systems are considered essential for pest management. Although there is no doubt that quarantine services can prevent the introduction of new pests into a country, the mere existence of a quarantine service is not a good indicator of effective prevention. Enforcement of quarantine regulations in developing countries is generally poor. Often the organizations responsible do not have the power, the technical means or the knowledge to detect the introduction of plant diseases efficiently.

Crop diversification.   In most developing countries, policies aimed at achieving crop diversification in farming systems are still the exception not the rule. Mostly, priority is given to cash crops because commodity-oriented institutions have been established and these are powerful political lobbyists. The result is large areas devoted to monoculture, which favours high populations of pests and encourages application of large amounts of pesticide.

Breeding resistant species.   This is one of the major policies that does not involve pesticides. The experience with rice in Asia, however, has shown that plant resistance alone is not a sufficient tool in pest management. Resistance to various pests and their biotypes tends to break down some time after the resistant variety is released. The mechanisms of this breakdown are not fully understood. Pesticides may play a role in this process through their negative impact on beneficial organisms. If the populations of these are reduced, the balance of the ecosystem is changed in favour of pests, which could accelerate the loss of resistance.

Pesticide policies

Pesticide policies are those that directly or indirectly influence the use of pesticides in a country. They can be assigned to three groups: regulatory policies, pricing policies and government investments.

Regulatory policies.   These command and control policies deal with factors other than the price which influence supply and demand of pesticides. One approach is to establish regulatory institutions responsible for registration, restriction or banning of certain chemicals. Even in industrialized countries these decisions are based on complicated and at the same time questionable risk-benefit analysis. Such analysis is supposed to judge whether a certain pesticide should be used in a given country on the basis of a comparison of its specific benefits and costs (risks). Regulatory policies are heavily influenced by the groups affected by their decisions: the pesticide industry, consumers, researchers, farm managers and farm labourers. The interests of these different groups have to be balanced by the regulatory institution. The choice of what particular risks and benefits will be considered and especially to what extent they will be evaluated is mostly being left to the agencies.

In theory, there are two ways in which policy decisions may be made. In the first, regulatory decision making is governed by who will gain and who will lose as a result of the decisions. The second way is that policies are determined

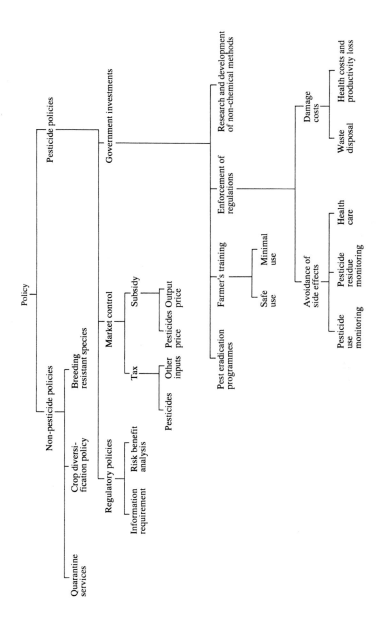

FIG. 2.   Types of agricultural policies related to national pest management.

by those whose interest is most effectively presented in agency deliberations and therefore who is likely to be most influential (von Ravenswaay & Skelding 1985). In practice, the various groups involved in decision making either assert political pressure or provide information which the regulating agency depends upon. Analysis of the different groups shows that industry, farm managers and researchers have very similar interests. This is because of the general interdependence between industry as a sponsor of research and industry as the major source of extension information for farmers. Also, these groups are comparatively small, which allows them to organize their interests effectively. Consumers, farm labourers and those who apply the pesticides are a large, diverse group; consequently, the costs of organizing themselves are high and it is therefore likely that their real interests are not adequately reflected in the decision-making process.

An analysis of pesticide regulatory decisions made by the US Environmental Protection Agency (EPA) showed that a chemical is most likely to be banned if it is harmful to those who apply it (Cropper et al 1992). Statistically, the death of one applier was equal to the deaths of 600 consumers. In developing countries, priorities are likely to be the opposite: cases of severe food contamination will be treated with high priority because they affect the rich groups of the society. Problems of occupational health and pesticide poisoning among farm labourers will receive less attention because they affect poor people: these events are often readily attributed to improper use of the pesticide. The many cases of pesticide poisoning documented in Thailand (Whangthongtham 1990) have so far not stimulated a strong government reaction.

Pricing policies. These deal with the price factors related to pesticide use, basically taxes and subsidies. Taxes can be import levies or local taxes. In most countries pesticides are subject to some kind of import tax. Although taxes are in principle the opposite of subsidies, a tax structure that favours pesticides relative to other inputs is a subsidy. In Thailand, pesticides are taxed at a low rate relative to other inputs such as fertilizers and agricultural machinery.

An overview of the possible subsidies is given in Table 1. Pesticide imports are sometimes given favourable exchange rates (Repetto 1985). The price policy for other inputs, especially fertilizer, can affect the demand for pesticides because of complementarity, i.e. if fertilizer consumption is stimulated by subsidized prices, pesticide demand will also increase.

Recently, direct price subsidies have been reduced or eliminated in many developing countries; however, non-price subsidies still exist (Waibel 1990).

The major subsidies, which pass largely unnoticed, are governments' emergency funds for counteracting pest outbreaks. Because of annual replenishment procedures, these funds are spent whether or not there has been a pest outbreak. For example, the government of Thailand assigns US$ 10 million annually to an emergency budget for pest outbreaks. This amounts to

TABLE 1   Types of pesticide subsidies

	Based on: *Price factors*	*Non-price factors*
Obvious	Government sells pesticides	Government's investments in reducing pesticide damage: residue laboratories;
	Government refunds pesticide companies costs	training on safe use of pesticides
	Subsidized credit for pesticides	Government's investments in pesticide research
	Preferential rates for tax and exchange rate	
Hidden	Government agency outbreak budget	Lack of pest definition
	Price subsidies of outputs	Wrong definition of crop loss
	Price subsidies of complementary inputs	Lack of data on crop loss
	Poor implementation of code of conduct	Lack of statistics on pesticide use
	Externalities of pesticide production and use	Lack of transparency in regulatory decision making
	Marginal user costs of pesticides	Training curricula of plant protection extension workers

a substantial share of the insecticide market for rice, because most of the support is given to rice farmers. Furthermore, the lack of government intervention in the pesticide market with its obvious negative externalities such as detriment of human health, must be considered as a hidden, indirect subsidy. The same is true for the general information environment associated with pesticide use.

Because assumed crop losses are often overestimated, pesticides are considered as indispensible production inputs. Pest eradication programmes are considered as a final solution to some pest problems and are still a frequent type of government intervention, despite their often questionable cost-effectiveness and obvious environmental damage. For example, in Thailand in 1989–1990 the government released US$ 20 million (not all for pesticides) for the eradication of the brown plant hopper. Eradication programmes are questionable environmentally and from the point of view of conserving genetic diversity.

Government investments.   The governments of developing countries are directly involved in the institutional use of pesticides, such as pest control campaigns to counteract outbreaks. A hidden type of institutionalized use of pesticides is their incorporation as a component of credit programmes. A survey in Costa

Rica (Thrupp 1990) showed that for most of the crops the input requirement in the credit guidelines is higher for pesticides (among which insecticides are often predominant) than for fertilizers. This includes crops such as rice, for which the use of insecticides and fungicides is known to be largely uneconomical.

Another type of government investment is support for research and development of non-chemical methods, such as biological means of pest control. Except for *Bacillus thuringiensis*, the results of such studies are not yet conclusive. The extent of these activities is still marginal compared with the support given to chemically based research.

Government investment also occurs through the enforcement of regulations directed at avoiding side-effects caused by pesticides. The costs of monitoring pesticides in food and water in industrialized countries are considerable. In Germany, an estimated DM 250 million per year would have to be spent on monitoring residues in water for it to be effective. This represents roughly 25% of the annual sales value of pesticides in Germany.

In developing countries the establishment of government laboratories to monitor pesticide residues has been supported by donor agencies. The cost-effectiveness of such monitoring for domestically consumed crops is questionable. Data on pesticide residues are rarely available to the public, because they are often considered confidential. Thus, investment in monitoring the side-effects of pesticides can become a constraint to IPM, as capital used in this way detracts from funds for methods that reduce reliance on pesticides, such as training programmes. Furthermore, because of the existing power structure in developing countries, there is a danger that data on chemical residues will be used to justify the continuation of a pesticide-oriented policy. Donor agencies therefore should consider stimulating high-income consumer groups to set-up privately run laboratories to monitor pesticide residues. In this case, collaborative projects with the rich classes of society in developing countries might be justified because they would help the poor farmers in the long term by stimulating the development of alternative technologies.

Finally, government investment in the disposal of unused pesticides may have the same effect as investments in monitoring side-effects as regards IPM. Such investments ignore the principle that the polluter should pay and are, strictly speaking, another type of subsidy. Yet in an emergency, as is the case in many African countries that have participated in locust control campaigns, these investments are justified on humanitarian grounds. These investments, however, should be accompanied by long-term strategies preventing such a situation.

Government investments often aim at minimizing or avoiding the side-effects of pesticides. Their opportunity costs are the benefits that could have obtained if the funds had been invested in programmes which reduce the reliance on pesticides, such as training. As shown by recent loss assessment studies (E. C. Oerke & F. Schoenbeck, unpublished paper, 48th Ger Conf Plant Prot, Göttingen, 5–8 October 1992), yield losses are rising despite increasing pesticide use. This means that the opportunity costs of the current emphasis of government investments are rising in relative terms, because the rates of returns from pesticide-based investments are decreasing. Furthermore, the scale of current investments in minimizing the side-effects of pesticides may be overestimated because existing levels of pesticide use may be too high in the first place.

Recommendations

This brief and certainly incomplete analysis of some important pest management-related policy measures in developing countries prompts some thoughts on further investigations. Firstly, we need more information on the economics of the present level of pesticide use in the major crops worldwide. We need more collaboration between economists and plant protection specialists to analyse trials in farmers' fields and to utilize pest population counts or estimates taken from untreated plots as inputs to economic analysis. The main element missing from most economic analyses of pest problems is some damage function showing the causual relationship between pest levels and crop yields. A comprehensive cost-benefit analysis should be conducted for various crops in collaboration with the relevant International Agricultural Research Centres. For accurate estimates of the benefits of pesticide use, there should be a crop-specific compilation of trials worldwide in which alternative treatments have been used. Basically, this would be the type of study carried out by Herdt et al (1984) but using data from farmers' fields instead of experimental stations.

In research, emphasis should be shifted away from the economic threshold of pests. Investments could be in an agroecosystems-type of training, like that practised in the Indonesian IPM project supported by the FAO. In this programme, farmers discover the relationships between beneficial organisms and pests through repeated field investigations. Field observations are summarized on pictures drawn by the farmers themselves; these are used as a basis for discussion and decisions about future actions. The Indonesian experience has shown that it is useful to determine research needs through participatory training. There should be a move away from supply-driven research to more demand-led or 'training-driven' studies.

As shown in Indonesia, however, training farmers may be effective only if it is accompanied by policy changes which favour IPM. Policy, therefore, has to be 'training supportive'.

The first requirement in setting up policies is planning. This requires data and information. However, these are largely monopolized by the pesticide industry. For example, in Germany it is not possible to obtain details of pesticide use by crop or farming system. Data of this kind on a worldwide basis are available from market research companies but at a high price. Nevertheless, an international donor consortium should not consider these costs as prohibitive: they could save large amounts of money currently spent on surveys in projects that are never compiled comprehensively. The results obtained could also support an urgently needed macro-level monitoring of the impact of pest management programmes.

The next step is to decide what kinds of pesticides are required for developing countries in the future. For example, one should ask whether sufficient evidence can be provided that the use of WHO class Ia and Ib pesticides in developing countries is justified.

Such efforts will also improve risk-benefit assessments for pesticide registration. At present, benefit assessment is based on the crop loss that occurs when use of a pesticide is cancelled, often assuming that there is no substitute. This leads to an overestimation of benefits. Even in the US, for only 11% of pesticides considered for cancellation by the EPA is there supporting information from measured yield data (von Ravenswaay & Skelding 1985). Most cases are based simply on expert opinions. It is essential to produce a clear picture of the benefits of pesticide use, if one wants to depart from emotional judgements based on risk data.

The responsibility of donor countries as regards national plant protection policies must be stressed. Any incentive that promotes the use of pesticides above the socially optimal level should be discouraged and there should be intensive studies prior to collaboration in the field of plant protection. A major concern is the outbreak budget of plant protection organizations, which reserves funds to supply chemicals in the event of a major pest infestation. There is hardly any justification for these emergency budgets, particularly if plant protection administrators at the same time complain that they do not have sufficient funds for training.

The cost-effectiveness of government investments for regulatory and monitoring institutions should be assessed carefully. Investments should aim at reducing the dependence on pesticides and their overall use instead of reducing the side-effects. In avoiding side-effects, priority should be given to improving the amount and quality of information on risks and benefits of pesticide use. Government investments must pay more attention to training following the principle of adult education, stressing an ecosystems approach. Training programmes must also consider the main interactions of the ecosystem with the larger farming system, including institutional aspects. Support for training should be conditional on disincentives for pesticide use. Donor countries therefore should review their policies with regards to financial assistance. Pesticide assistance must be accompanied by strategic plans aiming at long-term solutions.

In summary, economic policy considerations have to become an essential part of any plant protection and pest management planning. The pesticide policy of today must be based on two main pillars: improving human capital through training in the management of agroecosystems and a strong emphasis on environmental policy which ultimately means economic evaluation of the importance of beneficial organisms in pest management. Shifting attention away from chemically based solutions and stressing the role of agronomic measures in reducing the probability of pest occurrences should become a main concern of pest management policy in developing countries.

Acknowledgements

I would like to thank Professor Hartwig de Haen and Dr Peter Kenmore from the Food and Agricultural Organization of the United Nations, Dr Rolf Link and Dr Thomas Engelhardt from the Deutsche Gesellschaft für Technische Zusammenarbeit, and Josef Gamperl from the Kreditanstalt für Wiederaufbau for critical comments on earlier versions of this paper.

References

Anonymous 1990 Some growth in world market 1989. Agrow 118:16. Cited in: Knirsch J 1990 Lieferant für den Weltmarkt—Stellenwert, rechtliche Regelung und Bewertung der bundesdeutschen Pestizidausfuhren (Supplier to the world market—importance, legal regulations and evaluation of the German pesticide exports; in German). In: Ruhnau M, Altenburger R, Boedecker W (eds) Pestizid Report. Verlag Die Werkstatt, Göttingen, p 147–189

Asian Development Bank 1987 Handbook on the use of pesticides in the Asian Pacific Region, Manila, Philippines. Asian Development Bank, Manila

Cropper ML, Evans WN, Berardi S, Dulca-Soares U, Portney PR 1992 The determinants of pesticide regulation: a statistical analysis of EPA decision making. J Polit Econ 100:175–197

Hansen K 1990 The first three years: implementation of the World Bank pesticide guidelines 1985–1988. Consumer Policy Institute, Washington, DC

Herdt RW, Castillo LL, Jayasuriya SK 1984 The economics of insect control in the Philippines. Judicious and efficient use of insecticide. International Rice Research Institute, Los Banos, Philippines

Kenmore PE 1990 Indonesia's integrated pest mangement—a model for Asia. FAO Project Report. FAO, Manila

Knirsch J 1990 Lieferant für den Weltmarkt—Stellenwert, rechtliche Regelung und Bewertung der bundesdeutschen Pestizidausfuhren (Supplier to the world market—importance, legal regulations and evaluation of the German pesticide exports; in German). In: Ruhnau M, Altenburger R, Boedecker W (eds) Pestizid Report. Verlag Die Werkstatt, Göttingen, p 147–189

Pesticide Action Network (PAN) International 1989 The FAO code: missing ingredients. Prior informed consent in the international code of conduct on the distribution and use of pesticides. PAN International, London

Repetto R 1985 Paying the price: pesticide subsidies in developing countries. World Resources Institute, Washington, DC (World Res Inst Rep 2)

Rola A, Pingali PL 1992 Pesticides and rice productivity: an economic assessment for the Philippines. International Rice Research Institute, Los Banos, Philippines

Thrupp LA 1990 Inappropriate incentives for pesticide use: agricultural credit requirements in developing countries. Agriculture and Human Values. World Resources Institute, Washington, DC, p 62–69

Tüttinghoff H 1991 Pesticide use in plant protection. Descriptive analysis of decision making among rice farmers in central Thailand. Wissenschaftsverlag Vauk, Kiel (Farm Syst Res Econ Trop 10)

Vaagt G, Kern M 1992 Das Prior Informed Consent (PIC)—Verfahren als Hilfestellung
    für Entwicklungsländer. Mitt Biol Bundesanst Land- Forstwirtsch Berl-Dahlem 283:274
von Ravenswaay EO, Skelding PT 1985 The political economics of risk benefit assessment:
    the case of pesticides. Am J Agric Econ 12:971–976
Waibel H 1984 Die Wirtschaftlichkeit des chemischen Pflanzenschutzes im Reisanbau.
    PhD thesis, University of Hohenheim, Germany
Waibel H 1990 Pesticide subsidies and the diffusion of IPM in rice in southeast Asia:
    the case of Thailand. FAO Plant Prot Bull 38:105–111
Whangthomtham S 1990 Economic and environmental implications of two alternative
    citrus production systems. A case study from Pathum Thani Province in Thailand.
    MSc thesis, Asian Institute of Technology, Bangkok, Thailand
World Bank 1985 Guidelines for the selection and use of pesticides in Bank financed
    projects and their procurement when financed by the Bank. Cited in: Hansen K 1990
    The first three years—implementation of the World Bank pesticide guidelines
    1985–1988. Consumer Policy Institute, Washington, DC

DISCUSSION

Jayaraj: You mentioned pest and disease surveillance programmes. We have
been implementing fixed plot surveys and roving surveys for a few decades in
India. Even though the pest or disease outbreak may occur 500 m away from
a farmer's plot, monitoring gives a broad indication about the intensity and
spread of the problem and prepares the farmer to manage the situation.
However, there is a need for improved surveillance. We have been using new
tools in our surveillance programme, and it at least serves the purpose of
forewarning the farmers, even if we cannot accurately forecast pest outbreaks,
as the locust warning programmes can. Pest and disease surveillance programmes
are important, and essential for IPM. They should not be given up, they should
be given more emphasis.

Waibel: I agree surveillance systems have shown government researchers and
extension workers, especially in rice, what's going on in the field. The
surveillance systems have shown that actually there is not much happening. I
have done such work myself, looking for high population levels of pests, but
found them very rarely. My professor used to wish me good luck in the sense
of having higher pest populations! But surveillance systems cannot be taken
as a substitute for farmers' decision-making or for farmers' own observations.
This is where these programmes were wrongly designed. The idea was to have
at a central location a map with a lot of lights blinking once the reports came
from the fields, then the aeroplanes would start spraying. This didn't work.
Information feedback from the field is needed, but not in the way provided
by early surveillance systems.

Jayaraj: In 1986, herbicides accounted for 45% of the total world market
for pesticide use. In India, herbicides formed only 7% of the market. This may
be the case in other developing countries as well. Of course, we started using
herbicides much later. In developing countries, where farmers are mostly

illiterate, they need more training in the scientific use of herbicides. If herbicides are used unscientifically, the crop plants are also killed. Some chemicals are harmful to the intercrops or the succeeding crops, even in trace amounts. Atrazine kills all legumes planted as an intercrop or as a succeeding crop. If we want to promote the use of herbicides, we should have strong extension education programmes and training programmes. Then we will be able to manage weeds better and also the pests and diseases in the cropping systems.

Kenmore: Dr S. K. DeDatta, a long-time chair of Agronomy at IRRI, said he learned about preemergent use of rice herbicides from rice farmers in Central Luzon, Philippines. Researchers hadn't considered it appropriate for preplanting treatment because it was supposed to be used a bit later in the cropping schedule. Farmers tried it, got very quick phytotoxicity symptoms on their rice, and shifted back to an earlier application. They would flood the field with a quick flash flood, let the field dry and the weeds sprout, then they treated with herbicide. They did that 2–3 times and avoided using the herbicide after the crop had emerged, at least for the first 2–3 weeks.

So when the pest is a big one, like weeds (weeds have the biggest biomass of any pest), and the effect of bad application of the chemicals is clear, as in phytotoxicity to the rice crop, that kind of barrier-controlled research accelerates very quickly.

Nagarajan: I had an opportunity to discuss the use of weedicides with farmers. In north-western India, particularly in the Punjab and Haryana, where fodder is not a problem, farmers are willing to use weedicides. In other parts of India, weeds such as amaranthus and chenopodium are used as vegetables and as fodder for cattle. Therefore the farmers are reluctant to use weedicides because these plants are considered as a resource and not as a nuisance.

Varma: I agree that weedicide use on a large scale is difficult because in India the weeds are occasionally used as food plants. But in some parts of the country in some seasons, for example in north-west India in the *kharif* (wet season) crop, weeds are the major problem. A similar situation arises in other semi-arid parts of the world, like northern Nigeria. If weeds are controlled, very good yields are obtained. So in India, weedicides probably should have a greater share of the pesticide market than 7%. Weeds not only cause direct damage through competition, they also act as reservoirs for pathogens.

Zadoks: During a review in the rice polders in Kerala, we asked the farmers why they didn't use weedicides. They replied that the labour unions are very strong in Kerala. The labour unions want work—hand weeding. The unions threatened to burn the farmers' houses if hand weeding was eliminated. That is a very interesting approach to the use of weedicides!

Kenmore: A sort of a higher-order subsidy to pesticides is inherent in the university education system. This comes from your 'perceived' crop loss. At university or in technical courses, for the sake of convenience of the faculty, the students may be given one list of pests and a corresponding list of pesticides, because that is easier to test in examinations. This simple correspondence is what those people remember. An entire generation of students coming into crop protection systems will hold on to that knowledge, utterly inappropriately. It's important to feed information for the longer term future back into university and diploma technical school curricula. The concepts on IPM that the discussions here have revealed to be common belief, at least from all the crop protection scientists in this room, have to be put into the hands of the people who are actually faced with students every day. We must get away from this list of pests and accompanying pesticides.

Nagarajan: Professor Waibel mentioned a delightful definition: the socially optimal level of pesticide usage. It reminds me of the yield gap analysis that we did for many crops. The present pesticide usage can be ranked as the maximum unwanted level; the amount used in an IPM programme is probably the optimal level of pesticide usage. The socially optimal level of pesticide which is acceptable is at this stage very difficult to identify, because it is too early to set a target for the reduction in pesticide usage achievable by using IPM. But if we implement IPM for a while, whatever pesticide usage remains can be taken as the socially acceptable one.

Rabbinge: You have to distinguish between objectives and instruments. If you are formulating objectives, you should formulate them in their own terms. If you want to minimize pesticide use, you should express the objective in terms of pesticide use per unit product or per unit area. If you want to express social objectives, you should express them in terms of minimizing, for example, the decrease of labour in agriculture. This distinguishes the objectives from the instruments you are using.

There are different types of instruments, as you described. They can be economic instruments, such as levies or subsidies. They can also be social instruments, such as giving more information, stimulating innovation or stimulating the development of supervised control systems. If you express everything immediately in monetary terms, you rapidly face a very difficult problem—how to express crop losses and reduction of pesticide use in monetary terms. If you express each objective in its own terms, not in monetary terms, the aims become much more explicit and it is easier to compare the different objectives.

Upton: In measuring crop losses, whether in monetary terms or in physical terms, would it not be more appropriate to use a probabilistic approach (that is, to estimate the probabilities of different levels of crop loss) rather than to provide a single estimate? The uncertainty about crop losses may be another reason for excessive use of pesticides. If farmers are risk averse, they will use excess pesticide as an insurance policy to avoid risk.

Zadoks: We have analysed risk aversion in The Netherlands in a very primitive way. Depending upon the training and social conditions, farmers in one region were risk averse, in another region they were less concerned about risk. There was no other relevant difference between the two regions.

Waibel: This discussion shows that part of the cost of pesticide use today which is being ignored is the resource cost. Organisms, whether they are called pests or friends or other organisms, are resources. When we use pesticides, we are depleting these resources. One of the resources is susceptibility of the pest to a pesticide. I am not sure whether it is a renewable or non-renewable resource, but bringing this cost into the equation means that you would at least have to take care that the rate of pesticide use was less than that at which susceptibility is reduced. If susceptibility is a non-renewable resource, you should be careful how much you are depleting this. Depleting natural organisms is another type of cost.

One can express the productivity of pesticide as ton of crop per dollar of pesticide use and compare this in different countries. In Japan, they may produce 3 kg of rice per dollar of pesticide use; in China they may produce one ton per dollar of pesticide use. Taking these data from a country and a crop over time would allow us to calculate these resource costs. There would be some complications, such as price factors, attainable yield shifts and so on, but it could be done. We are trying to get these data, but the agrochemicals industries don't want to give us these data, although they have them. Even in Germany, the regulatory agency cannot say how much pesticide, in monetary terms or in terms of active ingredient, has been used in wheat over the last 20 years. This is a severe constraint in terms of access to information. It would be beneficial to companies, if they realized that a situation may be reached where the resource cost of using a pesticide will make it unprofitable to market.

General discussion I

Food crops versus cash crops

Jayaraj: In India at the moment, there is a serious debate concerning the shift in area under cultivation from food crops to commercial crops. It is causing a lot of anxiety in my country and this may be the case also in other countries. It is a question of generating income for the farmers. When there is an industrial requirement for the product, area under food crops is diverted to commercial crops. We should consider 'sustainable crop protection', which our Chairman coined earlier (Zadoks, this volume), from this point of view.

The area under cotton is increasing, and cotton is grown in overlapping seasons. A lot of pesticide is used on cotton. Although only about 5% of the cultivatable area in the whole country is planted with cotton, 55% of pesticide consumption is for cotton. So there are problems of pest resistance to pesticides, resurgence of target and non-target pests, residues in foods, etc. The same pests spread to grain legumes, other coarse grains and vegetables, creating more problems.

We have to develop strategies to maintain the balance among crops in a given area. We should provide support to the farmers to prevent this shift from food crops to commercial crops.

Dr Jones described different farming systems (Jones, this volume). Farming systems, particularly those involving livestock and aquaculture, will promote the concept of sustainable crop protection, because the residues from these occupations will help in organic farming. So for sustainable crop protection we have to promote the concept of farming systems.

Upton: It confuses the issue to link the food crop versus cash crop question with that of pest control. It is probably true that cash crops are more prone to pests, or require more careful pest control than food crops. Nonetheless, I prefer to see these treated as separate subjects. Certain crops may be particularly disease prone, and one could discuss whether these crops should be avoided as a means of pest control.

On the question of food versus cash crops, there has been a lot of, I feel, misguided debate. There is a suggestion that cash crop production is competing with food crops, and causing declines in food production. All the evidence shows that in the areas where cash crop production has increased, so too has food crop production. The crops are not really competitive in that sense. Furthermore, it's questionable whether it's a bad thing to promote cash crops. If they are profitable, what's wrong with producing them?

Wightman: It's a nice concept to tell farmers that they should grow cash crops so that they can buy food, but somebody still has to grow the food crops. Unfortunately, wherever you try to grow legumes, especially those that will be eaten in India, you are faced with the same set of pests and probably diseases.

Nagarajan: The area under commercial crops is primarily decided by the pricing policy—the relative price of cotton or sugar in the market. There was a marginal increase in the area under cotton because India was short of medium staple cotton: supplies of short staple and long staple were already overflowing.

The real diversion that has happened concerns at least three crops. In north-west India, barley has been almost totally replaced by mustard/rape seed. Another case is the competition that is gradually emerging with both soybean and sunflower. Sunflower will be a tough competitive crop for many of the major cereals, because of the present high oil seed prices. The third scenario that is likely to emerge is due to the high prices of pulses, particularly chick pea and pigeon pea. We anticipate that these crops will encroach on the area under wheat, which has fallen marginally in north-west India in the past five years.

Jayaraj: The area under sugarcane is increasing at the expense of rice in peninsular India.

Norton: I would like to consider how one might use the commercial sector in the implementation of IPM. In Australia, about 90% of the cotton crop is supervised in terms of pest management by consultants. Given that IPM requires more sophisticated management, which many farmers, because they are doing lots of other things, involved in other businesses perhaps, can't achieve, to what extent do you think consultants might become involved? Are they a factor in India? What is the potential for using consultants, perhaps at the village level, who would get some payment in kind?

Zadoks: A country like India could promote this type of development. There are so many graduates of agricultural colleges who need employment and women who want to re-enter agriculture after child-bearing. One advantage of this system is that if a private consultant does the wrong thing, the farmer will not go again to him or her, and so he or she is out of business. This is not the case for governmental consultants, who may be blamed, but will be paid. Private consultants are not blamed but are no longer paid, and that's far more effective.

Mehrotra: Cotton in India is a highly pesticide-intensive crop; nearly 40% of the cost of cultivation is due to pesticides (Seetha Prabhu 1989). Because land holdings are small, consultants are not working here; the cotton growers take advice mainly from pesticide dealers. A recent survey in Punjab found that only 6% of the farmers look to the agriculture university or the extension departments for advice. About 60% of the farmers look to the pesticide dealers who give them credit; about 20% of farmers take the advice of the dealers, whether they give credit or not. The remaining 10–15% of farmers simply follow what the neighbouring farmer is doing.

Recently, the pesticide industry has formed a pyrethroid efficacy group that advises farmers to use pesticides, especially pyrethroids, judiciously. This has had a significant effect on cotton cultivation. In 1980, cotton accounted for 55% of the total pesticide consumed in India (Kapadia & Mohala 1980). This fell to about 40.5% in 1989–1990 (Sengupta 1990), since when it has remained about the same (David 1992). The situation, as far as plant protection in cotton is concerned, is complicated because farmers grow far too many varieties which differ in their flowering times and number of pickings.

What is disturbing now is the tremendous increase of pesticide consumption in paddy. Paddy used to account for 15–17% of total pesticide use in India (Kapadia & Mohala 1980). This increased to nearly 23% by 1984–1985 (Srivastava & Patel 1990) and was nearly 30% by 1989–1990 (Sengupta 1990). The percentage of total pesticide use is only an indicator, but if one takes into account the annual growth in pesticide production in India, it becomes significant. At present, no paddy pest has pesticide resistance. I'm afraid that the increasing use of pesticides on paddy may lead to a situation where most of the paddy pests in India develop resistance. This is the danger.

Another problem is the cultivation of sunflower. Usually, sunflower, cotton and pulses are grown in the same area. This accentuates the problem of *Helicoverpa*. Pesticide resistance in *Helicoverpa* has already created problems in Andhra Pradesh, Tamil Nadu and Punjab (Mehrotra 1992, Mehrotra & Phokela 1991). Similarly, in the zeal to increase oil seed production, mustard has been promoted in a major way in South India. This has led to many pest and disease problems with mustard in the south that do not exist in the north.

Thus, there is a need to look at the whole cropping system in relation to pesticide use. Pesticide use now is at the complete discretion of the farmers.

Wightman: When talking about the area of cotton and associated crops, we can't really split off this particular process crop, because it's all tied up with the same farming system. The key issue, as Dr Mehrotra said, is *Helicoverpa*. It flits from the cotton to usually pigeon pea or chick pea and causes havoc. This is part of the reason pulse production in India is fairly stagnant.

Ragunathan: In the last 15 years, pesticide consumption in India has been increasing gradually. This is largely due to the fact that a greater area is being covered by plant protection measures. The consumption per hectare is declining.

The area of cultivation of different crops in India has not changed much in the last 30 years. However, the production levels of some crops, like sugarcane, wheat and rice, have increased enormously because of the introduction of high-yielding varieties and improved agronomic practices. Production levels for other crops, like oil seeds, pulses and cotton, have not increased by much because these crops are mostly grown under rain-fed conditions.

In recent years herbicide use has increased in intensively cropped areas. Herbicides constitute about 7% of the total pesticide consumed annually. Wherever herbicides are used continuously, new weed problems are emerging for which we have no solutions. For example in Punjab and Haryana, new weeds are posing problems in both rice and wheat.

The role of women in crop protection and sustainable agriculture

Jayaraj: In the last 7–8 years, in Kerala and in Tamil Nadu, the percentage of female students attending agricultural universities has increased dramatically. Today, 55–60% of those studying for a BSc (Agri) or BSc (Hort) are girls. When they go to the villages, they are able to interact with the women farmers better. In Tamil Nadu, we organized a development project for women in agriculture. The programme included the production of biocontrol agents, botanical pesticides, etc, and it was successful. So there is a healthy change taking place in the rural areas where more and more women are aware of the usefulness of IPM. Through trained women graduates from the farm universities, we may be able to organize IPM better in the coming years.

Hoon: Unfortunately, the women don't end up working as extension workers. They get married and then they are out of the picture. We have been thinking about providing jobs for women after they have finished bearing and rearing their children. When these women reach the age of about 38–40 and their children are grown up, we should bring them back into agriculture. I have rarely seen women extension workers working in the villages. This is a new phenomenon that will happen more in the future.

Zadoks: I have seen several women in extension in some parts of India, but I must confess that they all were young!

Mehrotra: An interesting fact is that almost all of the work in agriculture is done by women. Except for ploughing and spraying the crops, the men don't do anything. The man sits as a pest! Recently, it has been proposed to the Pesticide Appliances Manufacturers Association that they develop a small pesticide applicator that can be used by the women. If this happens, the men will do no work, simply stay behind and drink. We need not only more women extension workers, but for women to be protected from having to apply the pesticides. There should be a law against women applying pesticides. As it is, the amount of pesticide, especially DDT and HCH, in breast milk of lactating females is very high in Third World countries, including India (Jansen 1983, Mehrotra 1985a,b, Bouman et al 1990).

Escalada: There's a real concern that pregnant and lactating women should not be exposed to pesticides. At the 15th session of the FAO Panel of Experts on Integrated Pest Control, Peter Kenmore and I drafted a recommendation on gender issues in IPM. We stated that women, and often the children who accompany them, are more vulnerable than men to pesticide-associated health hazards because they generally work longer hours, have less access to adequate nutrition within households and are under physiological stress during pregnancy and lactation. We recommended that active measures should be taken and new methodologies developed to increase the participation of women in IPM research, development and training. These actions should identify and address barriers to women's participation in IPM programmes at the institutional, community and household levels.

Kenmore: A gender analysis was done in the Indonesian IPM programme, which involves roughly a quarter of a million farmers. In the next five years we expect to have about one million farmers go through intensive training, so it's a pretty big programme. At this point, no more than 4–5% of those farmers have been women. The issue was not the training method. The training method, when done with women's groups, worked extremely well. The women are very adept at adopting IPM and using it. The problem was that there were structural obstacles. The initial definition of who could be involved in a government sponsored programme said that participants had to be members of official, government-sanctioned farmers' groups. The membership of these groups is 98% male. On the basis of our gender analysis, we've been able to change the criteria for being allowed to attend the farmers' schools. They are not restricted to members of officially registered farmers' groups, but are open to the communities from which those groups are drawn. I would like to echo Vineeta Hoon's comment on changing the nature of employment in the extension system by being able to recapture women who have been trained. This is a structural change; it is susceptible to policy decisions.

There is also the point Dr Escalada made about the Thai situation: if you are going to involve women in training courses, those courses have to happen at a time when women can attend.

Jayaraj: The Government of India and the Indian Council of Agricultural Research have decided to establish a farm science centre in every district, with the emphasis on training women. Women farmers can be trained in IPM tactics through these systems. We have discussion groups for men farmers and women farmers. So things are improving slowly.

References

Bouman H, Coppen R, Reinecke A, Becker P 1990 Levels of DDT and metabolites in breast milk from Kwa-Zulu mothers after DDT application for malaria control. Bull WHO 68:761–768

David BV 1992 Pesticide industry in India. In: David BV (ed) Pest management and pesticides: Indian scenario. Namrutha Publications, Madras, p 225–250

Jansen A 1983 Chemical contaminants in human milk. Residue Rev 89:1–128

Jones MJ 1993 Sustainable agriculture: an explanation of a concept. In: Crop protection and sustainable agriculture. Wiley, Chichester (Ciba Found Symp 177) p 30–47

Kapadia Z, Mohala D 1980 Pesticide marketing in India. In: David BV (ed) Indian pesticide industry—facts and figures. Vishwas Publications, Bombay, p 151–155

Mehrotra K 1985a Use of DDT and its environmental effects in India. Proc Indian Natl Sci Acad Part B Biol Sci 51:162–184

Mehrotra KN 1985b Use of HCH (BHC) and its environmental effects in India. Proc Indian Natl Sci Acad Part B Biol Sci 581–595

Mehrotra K 1992 Pesticide resistance in insect pests—Indian scenario. In: David BV (ed) Pest management and pesticides: Indian scenario. Namrutha Publications, Madras, p 17–27

Mehrotra K, Phokela A 1992 Pyrethroid resistance in *Helicoverpa armigera* Hubner V. Response of populations in Punjab on cotton. Pestic Res J 4:59–61

Seetha Prabhu K 1987 Pesticide use in Indian agriculture. Himalayan Publishing House, Bombay

Srivastava K, Patel N 1990 Pesticide industry in India. Oxford & IBH Publishers, New Delhi

Sengupta D 1990 Future developments required in India in the field of pesticide technology. Pesticide annual 1989/90. Colour Publications, Bombay, p 1–9 & 12

Zadoks JC 1993 Crop protection: why and how. In: Crop protection and sustainable agriculture. Wiley, Chichester (Ciba Found Symp 177) p 48–60

Agricultural development paths and pest management: a pragmatic view of sustainability

Geoff Norton

Cooperative Research Centre for Tropical Pest Management, University of Queensland, Brisbane, Queensland 4072, Australia

Abstract. Historical profiles can be used to portray past pathways of agricultural development, the factors that affected pest status and the responses made by farmers in the form of pest management. Understanding the key factors affecting these historical developments is thought to be crucial for identifying likely future scenarios and associated opportunities and constraints for improving pest management. Evidence for this view is provided by four case studies: brassica pests in the United Kingdom; tsetse fly and trypanosomiasis management in The Gambia; rice pest management in the Lop-Buri area of Thailand; and pest management in dryland cotton in north-east Australia.

1993 Crop protection and sustainable agriculture. Wiley, Chichester (Ciba Foundation Symposium 177) p 100–115

Pests are one of the major factors affecting the form of cropping systems. In temperate as well as tropical agricultural systems, the use of fallow, crop rotation, multiple cropping, transplanting and other cultivation practices have contributed to the control of pests, including insects, pathogens and weeds. In subsistence agriculture, adaptation to pests has been a major factor locking farmers into a particular type of cropping and farming system; for example, the three-field system of spring wheat, winter wheat and fallow in the UK (for over 800 years) and transplanted rice (over 9000 years) in China.

In more recent times, several factors have caused a shift away from these sustainable yet static cropping systems. Initially, the need for greater productivity, associated with a raised demand for food from a growing population that was increasingly urbanized, was the major factor. Malthus' dire predictions have been avoided through increased production per hectare as well as increased production per unit labour.

This urbanization process has also led to the development of modern agricultural technology, such as breeding techniques, mechanization, the

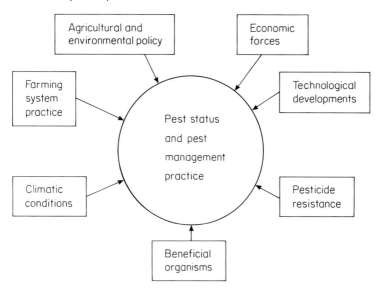

FIG. 1.   Factors influencing pest status and pest management.

production of chemical fertilizers and other inputs that raise production. Crop protection practices have also changed, as a result of changing pest problems faced by farmers, the new options for control available to them, and their changing cash and labour circumstances. More recently, regulatory issues such as agricultural protectionism and pesticide and environmental policies have all added to the complex of factors influencing this relationship (Fig. 1).

The significance of the historical view of pest management is not simply of academic interest. I would argue strongly that understanding the historical dimensions of pest problems is as important as understanding their immediate ecological, technical and socio-economic dimensions. The reason is that knowledge of the driving forces that have forged current pest management practices will give some idea of the forces likely to be acting in the future, affecting the status of pests and the feasibility of future pest management options. Decisions made on implementation, training, research and development need to be made in the light of most likely future scenarios, if these activities are to have significant and beneficial impacts.

This is the kernel of the idea I wish to present in this paper. I also wish to show how it can be put into practice, particularly in the context of pest problem definition workshops, involving scentists, extension officers, farmers and other key players. To elaborate this point, I shall describe four case studies, from the UK, The Gambia, Thailand and Australia.

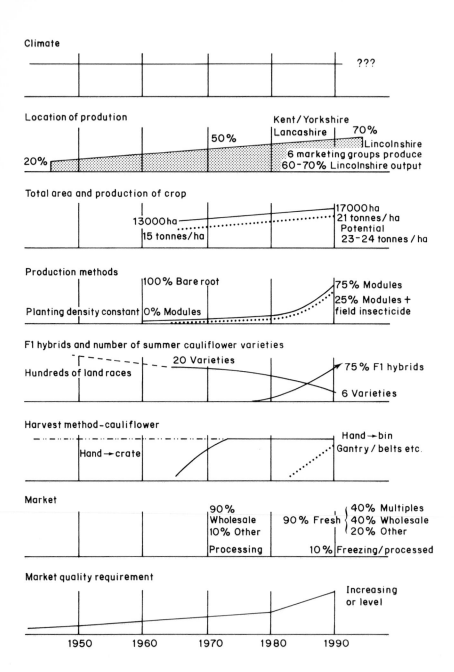

Climate

???

Location of prodution

20%   50%

Kent/Yorkshire
Lancashire   70%
Lincolnshire
6 marketing groups produce
60-70% Lincolnshire output

Total area and production of crop

13000ha
15 tonnes/ha

17000 ha
21 tonnes/ha
Potential
23-24 tonnes/ha

Production methods

Planting density constant

100% Bare root
0% Modules

75% Modules
25% Modules +
field insecticide

F1 hybrids and number of summer cauliflower varieties

Hundreds of land races

20 Varieties

75% F1 hybrids

6 Varieties

Harvest method-cauliflower

Hand→crate

Hand→bin
Gantry/belts etc.

Market

90%
Wholesale
10% Other

Processing

90% Fresh

40% Multiples
40% Wholesale
20% Other

10% Freezing/processed

Market quality requirement

Increasing
or level

1950   1960   1970   1980   1990

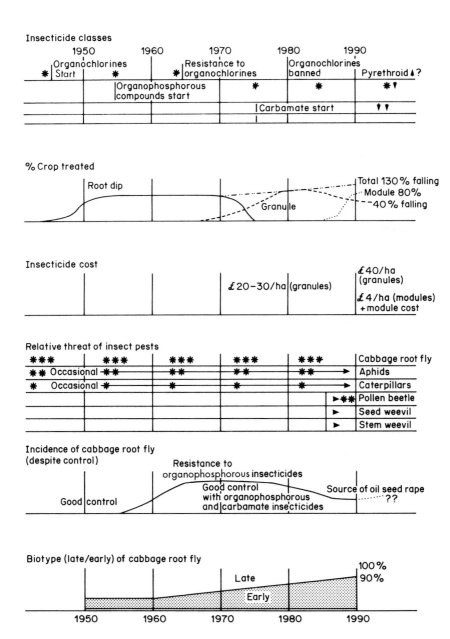

FIG. 2.   Historical profile for brassica pests in the UK.

Brassica pests in the UK

Cabbage root fly has been the major pest in brassica crops in the UK for many years. First-class research has been carried out on the biology and ecology of this pest, allowing a seasonal pest forecast to be produced. In 1990, a workshop was held to investigate how decision-making in brassica pest management (specifically in summer cauliflowers) might be improved.

During the workshop, the historical profile shown in Fig. 2 was constructed. It shows that many changes in cropping and marketing practice occurred between 1970 and 1990. The most significant in terms of cabbage root fly control was the rapid switch from transplanting plants with bare roots to transplanting plants in a modular peat block. An insecticide drench can be applied to the peat block, which reduces the amount of active ingredient used per hectare to 10% of that conventionally used in the form of granular application to the soil. This practice has also resulted in an equivalent reduction in the cost of insecticide per hectare. One conclusion is that any further reduction in insecticide use, as might be obtained from an early season forecast of cabbage root fly attack, for instance, is unlikely to be widely adopted, since growers now have a cheap and effective way of dealing with the problem.

Control of tsetse fly and trypanosomiasis in The Gambia

In March 1991 a half-day workshop was held at the Government Department of Livestock Services, Abuko, in The Gambia. The major objective of the workshop was to identify how past, present and future research on the tsetse fly–trypanosomiasis system, involving the indigenous N'Dama cattle, which are trypanotolerant, might best contribute to the information needs of the Department.

An historical profile that had been prepared at a previous half-day workshop was used to focus attention quickly on the key issues (Fig. 3). The highlights of the historical developments were confirmed by participants. Migration from rural to urban areas, especially in Banjul, means that the rural population is growing more slowly than the overall population. The cattle population seems to be stabilizing at about 300 000 head. The amount of land used for crops or fallow is increasing; more of this land is being fenced. Range management is becoming more intensive with more boreholes and internal fencing. The population of warthogs, an alternative host of tsetse fly, has declined because of loss of habitat. Tsetse fly abundance has fallen because of loss of suitable habitat, reductions in the populations of alternative hosts and reduced rainfall. Climatic models indicate that the downward trend in rainfall in recent years may continue. Human sleeping sickness is now virtually unknown and the prevalence of trypanosomiasis in cattle has declined to about 3%, but varies between districts—in 1990 it was 20% in Bansang and 1–2% in Kenaba.

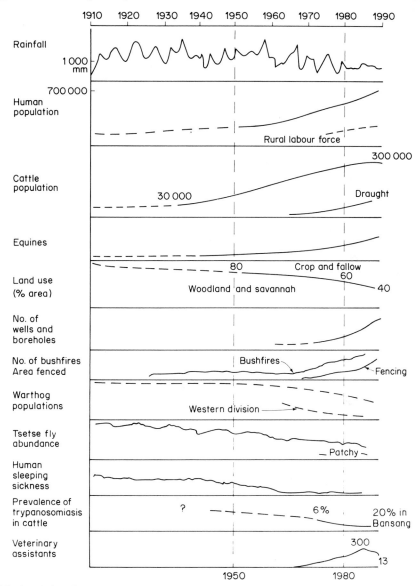

FIG. 3.   Historical profile for tsetse-trypanosomiasis in The Gambia.

Using this historical profile as a starting point, the participants in the workshop focused on the developments they thought likely to occur over the next five to ten years. This involved five main topics: land use and management; livestock developments (especially of cattle); equines; incidence of tsetse fly and trypanosomiasis and their control; and veterinary assistants.

In general, most of the trends are showing that the incidence of tsetse fly and trypanosomiasis in cattle in The Gambia is likely to decline in the future, particularly as the result of the increasing human population and the reduction in bush. However, other developments may reverse this, such as the increase in the number of (susceptible) equines. Changes in land use and management may also increase the risk of tsetse fly and trypanosomiasis. For instance, forest parks are likely to increase in the future, providing potential sites for tsetse fly. There will be increasing pressure to use these parks for cattle grazing.

On the basis of this systematic analysis of likely future scenarios, as facilitated by the historical profile, several recommendations were proposed by the workshop group, the majority being aimed at sustaining the trypanotolerance of the indigenous N'Dama cattle. Three of the more important recommendations are presented below.

Recommendation to utilize experience from Senegal. Monitor developments taking place in Senegal, their successes and failures, to obtain an early warning of the developments that might occur in The Gambia.

Recommendation to monitor shifts in cattle breeds. Surveys need to be undertaken to monitor changes in the genetic structure of cattle in The Gambia, particularly any trend in increasing Zebu crosses, which are far less tolerant to trypanosomiasis.

Recommendation on integrated farming and tsetse fly. A collaborative project, involving the Department of Livestock Services, the International Trypanotolerant Centre and the Livestock Owners Association should be started to investigate any increased risk of tsetse or trypanosomiasis resulting from deferred grazing (areas reserved to provide fodder in drought conditions) and forest grazing strategies associated with integrated farming schemes.

In this second case study, the major value of the historical profile has been twofold: to identify possible ways in which the sustainability of cattle and milk production, provided by the trypanotolerant N'Dama cattle, might be lost, and to determine the options for monitoring and preventing this.

Rice pest management in the Lop-Buri area of Thailand

The International Rice Research Institute IPM (integrated pest management) Research Network, together with respective national agricultural research systems, has conducted problem-based workshops in six countries in Asia over the past two years. The five-day Lop-Buri workshop in 1990 concentrated on the problem current at that time of brown plant hopper and ragged stunt virus, transmitted by the plant hopper. An historical profile of factors related to this problem was constructed during the workshop.

From this rough sketch, it was hypothesized that outbreaks of brown plant hopper and ragged stunt virus were related to the price of rice. When the price increased, because of world market supply and demand, Thai farmers switched to higher-yielding (more susceptible) varieties, they increased fertilizer use, increased the proportion of the second crop devoted to rice and increased insecticide use. The net result is an outbreak of brown plant hopper and ragged stunt virus. This hypothesis needs further investigation. More detailed information collated by Peter Kenmore (personal communication) certainly indicates a very strong positive relationship between insecticide use and numbers of the brown plant hopper.

This example shows how one can develop working hypotheses from an historical profile that could be useful in thinking about the options that might be taken to help prevent future outbreaks of the brown plant hopper/ragged stunt virus.

Pest management in dryland cotton in north-east Australia

The most recent workshop described here was carried out in Australia in September 1992. Its main purpose was to assess the problems likely to face growers of rain-grown cotton over the next five years and to determine how a new IPM project, recently funded by the Cotton Research and Development Corporation, will address these problems. Participants included growers, research scientists, crop consultants and representatives of the chemical industry.

Again, an historical profile provided a starting point for discussing future options for improving cotton pest management (Fig. 4). Changes in planting methods and particularly changes in the types of pesticides used have contributed to the altered pest status of mirids and aphids in recent years.

As well as producing detailed guidelines for the planned project and demonstration trial, an exciting outcome of one working group session was the initiation by cotton consultants of a proposal to run a 'transitional' IPM trial. Consultants already provide recommendations on cotton pest control on the basis of detailed monitoring and threshold decision rules. However, there are problems with pesticide resistance in *Helicoverpa* and increasing environmental pressures to reduce the amount of chemicals applied to cotton. Since the proposed research project to investigate the performance and feasibility of introduced biocontrol agents will take considerable time to reach a commercial stage, consultants were anxious to take a step towards the transition to a more biological programme of pest management.

Consequently, during the workshop, the consultants prepared a plan of action. This included higher threshold values than those currently used; when a threshold is exceeded, they will use 'the softest option currently available that does not involve an unacceptable commercial risk'. These trials will be conducted on ten to fifteen hectare plots in commercial growers' fields in the 1992/93 season.

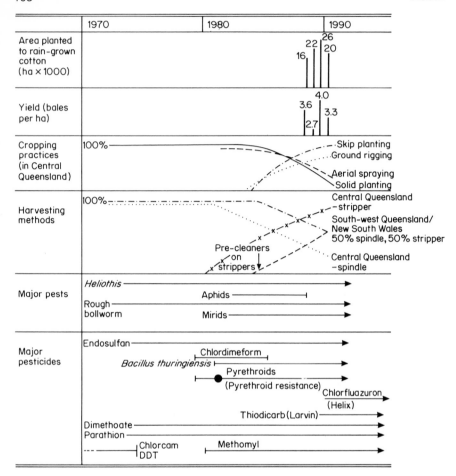

FIG. 4.   Historical profile for pest management in dryland cotton in north-east Australia.

Conclusions

This paper has attempted to show how the use of an historical profile in structured workshop sessions can provide a basis for rigorously defining current and future pest problems and for systematically exploring the options for improving pest management and sustaining agricultural production. This approach is a departure from the more usual scientific paradigms that dominate the field, particularly in entomology (Perkins 1982). The paradigms of 'chemical control', 'total pest management' and 'integrated pest management' are all solution based. The paradigm employed in this paper does not negate these other paradigms but provides an important, complementary viewpoint. The idea behind this problem-based paradigm is that before attempting to improve pest

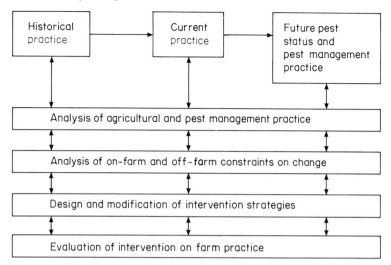

FIG. 5. An 'action research' approach to pest management.

management we need to have a clear, detailed, interdisciplinary definition of the problem, including a precise specification of how opportunities can be realized and constraints reduced.

As a methodology for achieving this, it is suggested that the 'action research' or 'soft systems' approach is adopted, a summary of this being provided in Fig. 5. Following problem analysis, including the use of historical profiles, a detailed study is made of opportunities and constraints to change. On the basis of this, intervention programmes are designed, involving research, training, policy and other appropriate measures. Once they have been implemented, the impact of these programmes is evaluated and on-course corrections made in an iterative manner. An important feature of this approach is that relevant players are involved as much as is feasible in the whole process, as illustrated in the case studies described above.

Acknowledgements

Many people have contributed to the ideas and information included in this paper, particularly those involved in the structured workshop sessions in the UK, The Gambia, Thailand and Australia. Although they are too numerous to name, I should like to extend my sincere thanks to them all.

Reference

Perkins JH 1982 Insects, experts, and the insecticide crisis: the quest for new pest management strategies. Plenum, New York

DISCUSSION

Royle: There seems to be a slight conflict in your definition of an historical profile. You have painted a picture where simple workshops generate all this information relatively easily. Yet the structure of your final paradigm seems rather complex. It would take quite a lot of time and effort to accumulate the information required in order to progress towards your goals. Am I right in interpreting it this way or not?

Norton: It depends on the situation. A two-hour workshop is better than no workshop, but clearly it is not all you need for every problem and more information needs to be collected. Therefore, one modifies the techniques and the amount of effort spent on the techniques according to the problem. It's amazing to me that quite often with these quick and simple methods you can very rapidly identify when research and development programmes are off course. But I wouldn't say they are always appropriate or useful.

Waibel: Having participated in a number of so-called 'logical framework workshops', where the same kind of problem analysis is being done in the preparation of projects in general, I often felt that one problem is the composition of the group. There are vested interests among the group members which cause the problem to be regarded in a certain way. This may not be a problem in your first case study, where the farmers were involved. There, the workshop members were really interested in the solution of the problem, because it was their problem. But this would be more problematic in the Lop-Buri case in Thailand.

Norton: I was using that particular example to illustrate how you can develop hypotheses, where in fact the price of rice seems to be closely related to brown plant hopper outbreak. Farmers were not involved in the Lop-Buri case, and there were considerable institutional problems. You have to set your objectives according to what's feasible within the workshop. In the Thai workshop we were trying to bring the extension and research scientists together to analyse the problem.

Neuenschwander: The German Gesellschaft für technische Zusammenarbeit (GTZ) has a very elaborate procedure to evaluate its projects: it's called ZOPP (meaning objectives-oriented project planning). GTZ runs all its projects through this procedure on several occasions: at the beginning, in the middle and at the end. In the few workshops where I participated, the major problem was exactly this question of who should participate. If you don't get the right people there, it is very difficult.

Jones: Geoff, this is very interesting. I don't want to sound too idealistic or critical, but I would like to ask, particularly with regard to The Gambia case study, what had that to do with sustainability? The historical profile you gave (Fig. 3) showed a number of trends, all of which were anti-sustainable—the population was increasing, the grazing was getting overgrazed, there was conflict over the water. You were solving one problem, that of trypanosomiasis, but what about the other problems and the interaction of your solutions with those problems? What's it going to be like in 10 years, let alone 100 years?

Norton: We solve some problems today, the world we tackle tomorrow! On that particular occasion, we were looking at a research programme on tsetse focused in The Gambia and seeing how we could apply the research results. We found that it wasn't a question of pest management as such, because the whole system has evolved under the pressure of tsetse, so there are trypanotolerant cattle. The point was that the tolerance could be lost for a number of reasons—if there was increased rainfall, if overgrazing causes the cattle to be moved to areas which have greater risk of tsetse fly. I agree other problems are arising. In a way, those are somebody else's problem. Can entomologists solve the world's problems? We were looking at only one aspect of sustainability, the threat of tsetse, which we tried to deal with as best we could, looking at the way in which the system was developing. You might not like the fact that the population is increasing, but are we going to get into human population control?

Mehrotra: The Indian Council of Agricultural Research organizes All India Coordinated Research Programmes for various crops, including cotton. The scientists working on these crops and a few subject specialists meet once or twice a year to assess the different problems occurring in the field. But nobody could have anticipated the magnitude of the build-up of resistance to pyrethroids in *Helicoverpa* that occurred in 1986. The 1987–1988 cotton season saw devastation of the crop in Andhra Pradesh by *Helicoverpa*, which received a lot of media attention. I agree that the composition and profile of the participants in the workshop is very important. The workshop should not only address the immediate problem but also consider what might happen in the future.

Norton: It's a necessary but not a sufficient condition to have the right people. You can get the right people together and they can go round and round in circles. The profile structures and focuses that discussion and you get fast results.

Rabbinge: I was interested to see that these historical profiles can reveal different developments and certain trends. But in three of the four examples, there were discontinuities. If you look at the timescale indicated, there were dramatic changes in methods or in other developments. In the cabbage case study, there was a complete change within four or five years of agricultural methods (Fig. 2). If you have a historical profile with these discontinuities, what can you then do with the trends? How can you use it to predict the future?

Norton: Let me give another example, rice growing in Malaysia, which has recently switched to direct seeding. All the developments there involve mechanization, which is driven by labour saving. So you know that people will be looking all the time for ways in which to save labour. If one had looked at the situation 10 years ago, one would have seen that the major labour-using activity left was transplanting. It accounted for probably half the man-hours involved in the crop. People were desperately looking at ways to reduce that labour, and they went to direct seeding. So the historical profile can give you a feel for the underlying driving forces.

Savary: Your presentation described a very clear pragmatic approach to problems, and it was enlightening in some aspects. You just mentioned the introduction of direct-seeded rice in Malaysia. This is occurring in many countries in South-East Asia where the labour force is declining. The fall in the labour force is a driving function for change and the solution is primarily mechanization, very careful control of the water supply and therefore very careful land preparation, and also the use of a lot of herbicides. So you are in fact dealing with a new crop, which has a lot more inputs and a lot more technology involved. It also has different pest problems. Could you elaborate on this change and its value in terms of sustainability of the system?

Norton: I agree it is a very different situation. It just means there has to be a tremendous change in research. When we conducted the first Malaysian workshop in 1982, we found that these developments were taking place but the information hadn't always got through to the research stations. Their research was still focused on transplanted rice. Most of the research that has been done on transplanted rice probably isn't applicable to direct-seeded rice.

Othman: The main weed problem in direct-seeded rice is grassy weeds. Good cultural practices (including land preparation, good selection of seeds and seed rates) and in-field water management are critical for weed control. We have carried out several programmes on direct seeding and its effects. If the water system for direct seeding is not efficient/good, weed control will be difficult; if the water problem is not solved, it will negate all efforts at chemical weed control. In Malaysia, about 90% of the rice is now planted by direct seeding. If the farmers cannot solve the problem of grassy weeds, they may shift to transplanting in the second season, and alternate direct seeding and transplanting in successive seasons.

Wightman: I like the preparation of historical profiles; I wish I had come across it 10–15 years ago. It seems to be similar to a technique ecologists and economists use, which is key factor analysis. If the data are available, you could do various kinds of arithmetic with it and formalize it a bit more.

Norton: That would defeat the object. If you get too deep, the analysis becomes an end in itself.

Wightman: The preparation of historical profiles seems a clear way of defining problems, if you have the information or access to the information that is required. Not all farming systems or regional data are available in this form. It seems this would be a major constraint in certain areas where I should like to try the same kind of procedure.

Norton: I agree that the detailed history may not be available. But often you find that someone who has been in the business for 20–25 years—usually extension officers, not research scientists—will just construct an historical profile straight off. We did one in Australia recently, where we had two groups developing profiles, with consultants in each group, and they came up with virtually identical profiles.

Escalada: Lack of detailed information is not necessarily a constraint. At the IPM Network workshop on rice leaf folder and brown plant hopper management in Vietnam, when participants were constructing historical profiles, they did not have all the relevant information with them. Historical data on the rainfall pattern, rice prices and cropping patterns in the Mekong Delta were not available at the time of the workshop. We asked the plant protection officer who was in charge to construct the profile when he returned to his regional centre and send it to us for integration with the workshop output.

In addition, we use historical profiles in conjunction with other diagnostic techniques, like decision profiles, seasonal profiles and discrimination profiles. Within these diagnostic workshops, we also conduct one-day farm surveys, just to get entomologists and plant pathologists to go out and talk to farmers. Most of the researchers we have dealt with in the workshops have not interviewed farmers before.

Battacharyya: For understanding of the present, knowledge about the past is very important. But in my experience, when we question the farmers about the past, they give only the good picture; normally they don't tell us about the bad experiences. We have to find a way to compensate for this.

Secondly, how do you differentiate between the felt or perceived need and the real need?

Thirdly, particular questions will suggest certain answers and you may miss the more important issues relating to the practice or the system.

Escalada: We have a correction for that. At the one-day diagnostic field interview we give the researchers a feel of how it is to talk to farmers and find out about their perceptions or knowledge level and current practice in pest management. Then the particular institutions involved in the workshop undertake their own large-scale survey after the workshop.

Norton: There are technical procedures for dealing with those sorts of problems. As Moni Escalada said, you try to get at the information from as many standpoints as possible. In Brazil, we did a quick survey of five farms in one day. We talked to extension agents. We then identified what we thought were the major trends in that particular case, and presented our results to another group of extension agents and got feedback from them. I agree you can be misled by farmers, so you have to ask as many people as possible, and eventually you get some consensus about what the real situation is.

Wightman: You concluded by proposing the way in which agricultural scientific projects should evolve. You are saying that scientists should sort out the best horse for the particular course. In some conditions, straight-forward biological control is appropriate and the best approach. In other systems, you have to consider whether you should be looking at all four or five components of potential IPM programmes for an individual case. If constructing a historical profile can help you make this decision, it is another strength of the method.

Kenmore: You described a problem-based approach: that's the direction in which scientists should go. There's another implication in what you have described, which is the question of accountability. To whom are the scientists responsible for the consequences of their recommendations? Involving the right people in the workshop is critical, but you also have to set up structures for future accountability. Follow-up is an extremely important part of the design, particularly in developing countries. Setting up ways for the people who practise agricultural science to become more accountable to farmers is at the core of what we are trying to do.

Norton: I couldn't agree more. In Australia, much of our funding comes from rural industry funds, so we have to be accountable. If we actually have an impact on the industry, we are likely to get more funds. Donor-funded projects don't have any of those built-in mechanisms: it's in everybody's interest to say that projects funded by donors are successful. The donors want it to be successful, the people doing it want it to be successful, the consultants coming in, who are paid by the donors, want it to be successful and so on. This is a real problem.

Upton: Geoff, the workshops you described all seemed to have reached fairly clear conclusions. Have instances arisen where there were several similar alternatives you needed to assess, and was there a need to do some budgeting or cost-benefit analysis of some kind?

Norton: Yes. One thing we find extremely useful is to use very simple models—economic models, investment models and so on, just to get a rough idea of what's going on in terms of cost benefit. In some workshops we did in Australia, we were looking at the problem of feral pigs and foot-and-mouth disease. There were various strategies: should you spend more money on quarantine or on monitoring the population of pigs, knowing that quarantine isn't 100% successful? You could hypothesize relationships between the degree of monitoring and the probable amount of time required for foot-and-mouth disease to be spotted once it has been introduced. There is also a relationship between the time taken to detect the disease, once introduced, and the cost of eradicating it. We hypothesized these relationships to get a structure, and then we could start looking at ecological relationships which would affect those cost functions. We found such a qualitative assessment much more useful than detailed quantitative cost benefits.

Escalada: At the workshop on golden snail management in the Philippines, John Mumford taught the participants to develop product discrimination profiles to rank farmers' choices of widely used control measures for snails, e.g. chemicals, hand-picking, transplanting older seedlings, installing screens across the irrigation water inlets onto their fields, pasturing ducks, water management and increasing plant density. A range of important attributes of ten control measures was determined from the opinions of participants and from comments made by farmers in the story videos prepared before the workshop. Chemicals were considered the most favourable because of ease of use, familiarity or simplicity, availability, reliability and effectiveness. However, when molluscicides were assessed in terms of safety and cost, they were ranked negatively. The profile presented a number of challenges to the introduction of non-chemical means of snail control. Developing a product discrimination profile was a qualitative exercise to determine what appeared to be the most popular options for farmers.

Kenmore: In field work by J. W. Ketelaar, on the golden snail in the same area, it was found that one very strong interacting variable is the position in the irrigation system. At the tail end of the irrigation system, nothing controls snails. The water flow concentrates the snails while water accumulation softens the rice to make it more susceptible. At the head end of the irrigation system, basically anything you do controls the snails because the population pressure isn't as high. This information came from field work with farmers. The magnitude of the damage at the tail end of the irrigation system was about 20 000 pesos per hectare, versus something under 2000 pesos per hectare at the head end. A good picture of what's happening in the field is very important.

Systems of plant protection

Paul S. Teng*, Serge Savary† and Imelda Revilla*

*International Rice Research Institute, Division of Plant Pathology, PO Box 933, 1099 Manila, the Philippines and †ORSTOM, Institut Français de Recherche Scientifique pour le Développement en Coopération, Centre ORSTOM de Montpellier, 911 avenue Agropolis, 34032 Montpellier Cedex, France

Abstract. A framework for associating pest management with farming systems may be characterized by a cropping index (number of crop cycles per year); a plant species homogeneity rank, described as the degree of monoculture or polyculture and intensity of use of modern, high-yielding varieties versus land races or traditional varieties; intensity of use of synthetic agrochemicals; the availability of societal infrastructure to support agriculture, such as irrigation and roads; and availability of support services such as credit, public/private sector research and extension services. Traditional methods of plant protection in the tribal areas of the Philippines rely heavily on ritual-based cultural practices that recognize the crop calendar and its activities as part of a larger social agenda. Most groundnut growers in West Africa utilize minimal inputs with little infrastructural support but they have often adopted improved varieties. Extensive farming systems with modern inputs are exemplified by wheat in the mid-western USA, where pests are managed mainly through host plant resistance. In intensive farming systems—vegetables in South-East Asia, wheat in The Netherlands and orchards in the USA—crop intensification is greatest and pest problems arising from overuse of pesticides are most noticeable. Unsustainable farming systems evolve if realistic plant protection is not taken into consideration.

1993 Crop protection and sustainable agriculture. Wiley, Chichester (Ciba Foundation symposium 177) p 116–139

Plant protection must be considered an integral part of farming to avoid the pitfalls caused by dealing with pests using techniques that do not recognize other components of the agroecosystem. Plant protection systems have co-evolved with farming systems; it is possible to identify patterns associated with the successful development or adoption of new pest management practices by farmers in the same way that one sees how patterns of farming systems have been designed in relation to landscape (Mollison 1988). These patterns, *inter alia*, have been influenced by features such as cropping intensity and yield goals, which in turn have been modified by population needs. In this paper, we shall examine how the co-evolution has occurred, using selected food-producing systems as examples. We shall also use an analysis of such systems to make suggestions for the development of sustainable plant protection practices that incorporate the pre-modern technology knowledge base in such systems.

A framework for analysing plant protection systems

Pest management may be associated with farming systems using a framework in which crop intensification is a key attribute (Table 1). The framework is characterized by several features, each assigned a value of 0–3. (1) A cropping index, which represents the number of crop cycles per year. (2) An estimate of the relative homogeneity of plant species, which is analogous to biological diversity. A reduction in diversity, owing to a high degree of monoculture or polyculture and intensive use of modern, high-yielding varieties rather than land races or traditional varieties, would have a high value of three. (3) Intensity of use of synthetic agrochemicals. (4) The availability of societal infrastructure to support agriculture, such as irrigation systems and roads. (5) Availability of support services such as credit, public/private sector research and extension services.

Other attributes are the extent of use and the effectiveness of plant protection techniques (represented as a comparative index in the lower part of Table 1). The values were added to give total scores that were used to compare the different cropping systems.

Within this framework, farming systems from different parts of the world may be considered as examples of plant protection systems. Traditional systems and ones based on indigenous knowledge, such as those in the tribal areas of the Philippines, rely heavily on cultural practices associated with rituals that recognize the crop calendar and related activities as part of a larger social agenda. Groundnut growers in West Africa utilize minimal inputs but have adopted improved varieties. In this situation, with potentially low yields and little infrastructural support, an economic objective may be difficult to discern. These systems are also inherently mixed with cotton and maize. Extensive farming systems which utilize modern inputs are exemplified by wheat-growing in the mid-western USA, where pest damage is controlled mainly through growing resistant varieties of host plants, and where low economic returns are still a disincentive to practise any plant protection during the crop season. Even though grain yields are lower than in the wheat-growing areas of Europe and Australasia, this area is consistently a major exporter of grain to the rest of the world. Intensive farming systems are exemplified by systems as diverse as vegetable growing in South-East Asia, irrigated rice production in the Mekong Delta, Vietnam, and sugar beet growing in the mid-western USA. In these systems, crop intensification (as defined above) is greatest and pest problems arising from overuse of pesticides are most noticeable.

Indices for the intensification of cropping systems have been proposed previously (Cox & Atkins 1979, Mollison 1988) but there is as yet little empirical study of the relationship between intensification and reduction of yields caused by pests. The study by Andow & Hidaka (1989) is one of the few.

TABLE 1 Comparison of cropping systems with respect to intensification parameters and plant protection techniques

	Traditional rice-based systems (Philippines)	Maize–rice–groundnut system (W. Africa)	Cotton–maize–groundnut system (W. Africa)	Wheat monocrop (mid-western USA)	Sugar beet (mid-western USA)	Rice in Mekong Delta (Vietnam)
Intensification parameters[a]						
Cropping index	1	1	2	2	1	3
Plant species homogeneity	1	1	1.5	3	3	3
Use of synthetic agrochemicals	0.5	0.5	1.5	1.5	3	3
Physical infrastructure	0.5	0.5	2	3	3	3
Support services	0.5	0	2	3	3	3
Total	3.5	5	9	12.5	13	15
Plant protection techniques[b]						
Host plant resistance	1/3	0.5/1	1.5/1.5	3/3	3/1	2/1
Pesticides	0.5/0.5	0/NA	2.5/1.5	0.5/0.5	3/0.5	3/1
Cultural methods	3/3	1.5/2	2/1.5	0.5/0.5	1/1	1/0.5
Natural/biological control	1/3	1/2	0.5/0.5	0/0	0/0	1/1
Rituals, others	3/1	3/1	2/1	0/0	0/0	1/1
Total	8.5/10.5	5.6/6	8.5/6	4/4	7/2.5	8/4.5

[a]Values 0–3; a low value represents a low degree of intensification. NA, not applicable.
[b]The first figure is an assessment of extent of use; the second figure is an assessment of effectiveness.

Plant protection in relation to cropping system

Traditional/indigenous rice cultivation in the Philippines

Rice-based tribal agricultural systems in the Philippines are scattered throughout the archipelago. From a survey conducted over two years, Teng et al (1991) reported that pest prevention or control practices adopted by traditional agriculturalists among the major ethnic tribes in the Philippines may be classified into eleven components: (1) timely planting of crops; (2) planting of border plants; (3) interplanting of resistant and susceptible rice varieties with definite spatial distribution; (4) continuous mixed cropping; (5) use of indigenous pesticides; (6) use of attractants to concentrate insect pests to facilitate pest control or to create a diversionary effect; (7) smoking out of insect pests with or without indigenous repellent agents; (8) biological control; (9) use of inert materials for pest control; (10) clean culture; and (11) use of mechanical traps. Interspersed among these practices are rituals which fulfil a social function and yet may have actual value in repelling pests. An example is the burning in the evening of *Imperata cylindrica* (a rice weed) in the belief that it repels rice pests. The ritual has three outcomes: the control of a rice weed, the management of insect pests of rice and the repulsion of mosquitoes. Cultural practices are also common (a cultural practice is any farm operation that will make the environment less favourable for pests to develop or multiply but which still favours crop production), although their use is often due more to necessity than intention because of the lack of resources of most of the farmers. Furthermore, farmers in such systems regard pests as part of the agricultural environment and consider pest losses of up to 20% 'normal' to the production system. The role of spatial variability in planting crops, although not recognized by farmers, is nevertheless part of the common practice. Traditional systems in the Philippines almost exclusively have poor or no access to infrastructural support and support services.

Biological control has arisen from the most subtle means devised by farmers in tribal systems to circumvent religious restrictions about killing other life forms. This awareness has become part of the legends and folklores transmitted orally from one generation to the next. For example, the knowledge of early Igorots of the activities of cats and hawks in controlling birds is integrated into a legend tracing the origin of these animals. Ethnic tribes such as the Mandayas, Agusanons, Batangans and Ibanags know the roles of the numerous species of birds, frogs, toads, lizards, snakes, spiders, wasps, ants and some small mammals in regulating pest populations. For example, the Batangans are aware that the plants in the immediate vicinities of their swiddens (temporary agricultural plots produced by cutting and burning off vegetable cover) or in the surrounding forest provide habitat for natural predators like the wasp ('mangans'), dragonfly ('sipsibud'), ants, spiders and birds. They seem to know that these habitats guarantee the continued presence of such natural enemies.

Spiders, dragonflies and a variety of smaller birds were particularly effective in controlling the adult populations of stem borers and corn borers ('tiklaw'). The wasps ('mangans') were reported by farmers to prey on insect larvae while the ants were noted to feed on insect eggs and less frequently on larvae.

The traditional maize–rice–groundnut cropping system in West Africa

In the savannah regions of West Africa, a typical traditional cropping system is represented by maize, upland rice and groundnut crops. Because there are sometimes very long fallow periods, the cropping index is relatively low, with a high species diversity. The system includes cattle, which are important in providing manure so that little chemical fertilizer is needed. The physical and support structures are very limited. Pest control measures include growing crops with partial resistance or tolerance to diseases, insects and weeds (which are essentially unrecognized). In practice, pest management is an inherent, non-specific component of the cropping practices, with a major emphasis on weed control and, to some extent, insects, through burning and tillage. Management of this cropping system—including pests—involves a very high degree of ritual practice.

The cotton–maize–groundnut cropping system in West Africa

The introduction of cotton induced this traditional subsistence farming system to evolve towards a more intensive one, where the export crop is predominant. Fallow periods are shortened and the cropping index is higher. Although components of the initial system, maize and groundnut, are still present, the species homogeneity is reduced. There is also a marked increase in synthetic agrochemical inputs (especially fertilizers, insecticides and herbicides), and enhanced physical and support infrastructures. In terms of plant protection, the system incorporates a fair amount of resistance to insects and diseases (e.g. in maize and cotton) and relies heavily on pesticides (insecticides on cotton), at the expense of natural mechanisms that stabilize the ecosystem. Improved land preparation, which is mechanized in about half of the villages, also probably contributes to some pest control.

Sugar beet in the mid-western USA

Sugar beet is a major source of sucrose and molasses in continental USA. An important beet-growing area is the Red River valley of Minnesota and North Dakota, where the warm summers and cool autumns are ideal for sugar production. The beet is grown as a monocrop with intensive use of inputs, especially fungicides against *Cercospora* leafspot, a fungal disease. In the 1970s, high-yielding varieties with relatively little resistance to the disease were grown over almost the whole area and there was reliance on benzimidazole fungicides. In 1981 severe epidemics caused up to about 50% crop loss because of widespread failure of the fungicides owing to resistant strains of the fungus (Shane et al 1985). This was a classic example of how unilateral pest control gave the impetus to develop systems of integrated pest management (IPM). Subsequently, research led to the deployment of a management system which used lower-yielding beet varieties and non-systemic fungicides applied with guidance from a weather-based forecasting system and knowledge of disease/loss relationships (Shane & Teng 1992).

Irrigated rice systems in the Mekong Delta, Vietnam

The Mekong Delta is probably the most intensively cropped rice area in South-East Asia, with up to three crops per year grown under a heavy regime of artificial fertilizers and synthetic pesticides. Outbreaks of the brown plant hopper caused severe losses in infested fields, showing an increase from 1986 to 1991 that corresponded to increasing use of broad-spectrum insecticides. In the same region, the application of fungicides to control blast and sheath blight diseases is common (Thuy & Thieu 1992). Recently, a leaf-yellowing syndrome was observed over several thousand hectares of the most intensively cropped districts and we have found a relationship of this syndrome to nitrogen use and soil condition (P. S. Teng, K. G. Cassman, T. W. Mew, unpublished 1992). The development of integrated crop protection systems for areas such as this is crucial for sustaining total food production but research is needed on methods that will also enhance the natural resource base (physical and biological).

Vegetables in South-East Asia

Vegetables, especially crucifers, are grown commercially under intensive conditions which favour the abuse of inputs, leading to breakdown of any natural system of pest control. All crucifers are susceptible to a large number of insects. The diamondback moth causes most concern because it has become resistant to nearly all insecticides used against it, starting with DDT (dichloro-diphenyltrichloroethane) in 1953 (Thalekar 1992). The great concern in this region of the world is that although regulations exist to govern types and timing of agrochemical treatments, in practice enforcement is weak and real threats to human health are posed by the high levels of pesticide residues in the vegetables sold to consumers. Host plant resistance has not been effective against the diamondback moth. Recently, combining the selective use of the biological insecticide, *Bacillus thuringiensis*, with release of a parasitoid appears successful in several countries.

Wheat in the mid-western USA

Wheat is grown under extensive conditions in the prairie region of the USA. The system is intensive in so far as a single crop is commonly grown with limited rotation and general use of fertilizer. Almost no pesticides are applied and reliance on host plant resistance to control diseases, especially the rusts, is standard practice. This dependence on resistant plants has worked because of an implicit gene rotation system owing to new releases of wheat varieties with different sources of resistance to the rusts, but also because of the geographic isolation in a climate that does not favour many generations of the fungus in each season. The severe winter in states such as Minnesota and North Dakota also contributes to reducing or eliminating sources of endogenous inoculum each season.

General issues affecting plant protection systems

Crop intensification and changing risk in agricultural production

Sustainability of agricultural production should be seen not only in terms of maintaining both the resources and outputs of a system, but also in the perspective of increasing production (Wilken 1991). This is especially true in developing countries, where the demand for food is expected to increase dramatically in the future. It is thus appropriate to discuss the consequences of production intensification from the perspective of plant protection.

Crop intensification may be considered as a gradual process, where cropping practices are altered or new practices introduced one at a time. One measure of risk, which might preferably be called risk magnitude (Rowe 1980), is crop loss. Each step in the intensification process is associated with a change in risk. There are relatively few references on the risk assessment associated with a whole intensification process. One example is peanut in West Africa, where a step-wise scheme for possible intensification has been delineated: (1) weed control, (2) increase of sowing rate and (3) increase in fertilizer input. The risk associated with this scheme has been measured experimentally. It was found that the multiple pathogens associated with peanut, especially peanut rust, were increasing crop losses much faster than the successive inputs were increasing yields (Fig. 1) (Savary & Brissot 1990).

The effects of some specific components of crop intensification have been documented in detail. For instance, an increase in the cropping intensity of potato in the Andean region of Peru cannot be considered without strict control of diseases caused by the nematodes *Globodera pallida* and *G. rostochiensis* (Glass & Thurston 1978). The effects of nitrogen fertilizer on pests and diseases of rice (Reddy et al 1979) and wheat (Daamen 1990) have also been extensively documented.

It is a common perception, largely verified by facts, that an increase of attainable yields (*sensu* Zadoks 1981) is associated with a corresponding increase

Yield   (kg / ha)

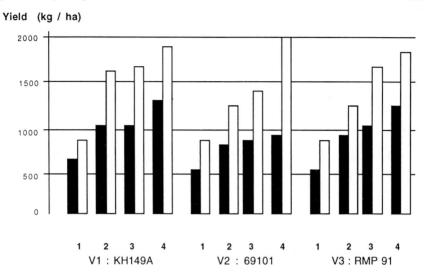

1   2   3   4   1   2   3   4   1   2   3   4
V1 : KH149A   V2 : 69101   V3 : RMP 91

FIG. 1.   The relationship between groundnut yield and diseases (leaf spots and rusts) in an intensification scheme in northern Ivory Coast. Open bars, average yields of plots sprayed with fungicide; closed bars; yields of plots not protected against foliar diseases. V1, V2 and V3 are three different durations of crop cycle. (1) Control; (2) weed control; (3) weed control + increased seeding rate; (4) weed control + increased seeding rate + fertilizer.

in crop losses (Cook & Veseth 1991). A crop loss database, containing components of cropping practices and components of the many pests in rice, is being developed at the International Rice Research Institute (IRRI). This should allow the issue to be addressed systematically. The database is compiled from the results of a series of complementary experiments, each involving a selected set of cropping practices and pest combinations. A partial summary of the available information in this database is given in Table 2.

Table 2 shows the bivariate frequency distribution of yields (Y) and attainable yields (Ya). Both yields are represented by a series of levels, from low to high. The levels were devised such that all levels would be represented by a reasonable number of plots (so that a chi-square statistic can be computed). The range of actual yield increases from two classes (Y1 and Y2) to several (Y2 to Y5), as the attainable yield increases progressively. In other words, the variability of actual yields increases with increasing attainable yields. This suggests therefore that the relationship between attainable and actual yield is not a straightforward one that could be represented by a simple curvilinear function.

Aside from the general relationship between attainable yield, which represents the yield levels achieved at each step of the intensification process, and actual yield, which represents the response of the system, including its pest components, special situations need to be considered. For rice, low to moderate crop losses

TABLE 2   A bivariate frequency distribution of plots using yield and attainable yield as classification variables

Classes of actual yield	Classes of attainable yield			
	Ya2	Ya3	Ya4	Ya5
Y1	23[a]	1	1	0
Y2	17	25	11	9
Y3	0	22	18	17
Y4	0	3	15	6
Y5	0	0	3	24

Yield categories in t/ha: 1, $0 \leqslant Y < 3.06$; 2, $3.06 \leqslant Y < 4.47$; 3, $4.47 \leqslant Y < 5.46$; 4, $5.46 \leqslant Y < 6.10$; 5, $Y \geqslant 6.10$.
[a]Entries are plot numbers.

are predominantly associated with low or moderate attainable yield (Fig. 2). However, when high or very high attainable yields are considered, two broad categories are predominant: negligible or low, and very high crop loss. In this case, it seems that high risk (crop loss) associated with high attainable yield corresponds to some pests or pest combinations, whereas other pests or pest combinations are not conducive to high risk.

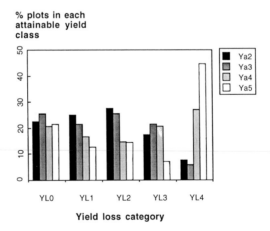

FIG. 2.   A rice crop loss database: distribution of crop loss categories as a function of four levels of attainable yield in rice. Five categories of yield loss (0, $0 > Y \geqslant 0.5$, $0.5 > Y \geqslant 1$, $1 > Y \geqslant 1.5$, and $Y > 1.5 \, t \, ha^{-1}$; YL0 to YL4) are considered. These categories are associated with different patterns of attainable yield (Ya1 to Ya5). The yield (attainable Ya and actual Y) categories are the same as in Table 2. There is no attainable yield categorized as 'very low' (Ya1). The five bars representing an attainable yield category add up to 100%.

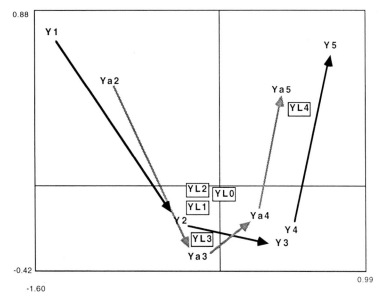

FIG. 3. A rice crop loss database: correspondence analysis of the relationships between the actual (Y) and attainable (Ya) yields and crop losses (YL). The diagram represents the various categories plotted using the two first axes (eigen vectors) generated by correspondence analyses. Paths of increasing actual (Y) and attainable (Ya) yields are indicated (Table 2). The analysis involved 195 plots.

A correspondence analysis may summarize the relationships between attainable yield, actual yield and crop loss in this database for rice (Fig. 3). It shows, of course, a very strong association between attainable and actual yield. But it also indicates that high (YL3) and very high (YL4) risks (crop losses) are triggered only when medium, high and, most especially, very high (Ya5) attainable yields are targeted. This diagram highlights the special relationships between very high attainable yield and very high crop losses. It should be noted, however, that neglible crop losses, being located at the centre of the graph, are associated with any of the considered attainable yields.

The situation of rice, which is affected by a range of pest combinations, contrasts with that of peanut, for which there are two main diseases (leaf spot and rust) in West Africa. In the latter case, a threshold of attainable yield was identified beyond which crop protection becomes a prerequisite (Savary & Zadoks 1992). As this threshold of attainable yield (1400–2300 kg ha^{-1}) roughly corresponds to the uppermost range of yield currently achieved in farmers' fields—where no specific disease control measures (e.g. varieties resistant to diseases and/or fungicides) are applied—the question arises whether crop intensification would be sustainable.

The discussion so far has emphasized the consequences of crop intensification and potential reduction of yields owing to diseases and insects, using crop loss as a measure of risk magnitude, rather than the occurrence or development of pests *per se*. A parallel can be drawn between the two main food crops worldwide, wheat and rice. In both cases, intensification relies heavily on the use of short-strawed varieties with a high harvest index; these varieties are able to sustain high yields through high inputs of fertilizer, especially nitrogen. Application of fertilizer increases the nitrogen content of plant tissues; short-strawed plants are closer to sources of disease inoculum; dense stands with a high leaf area index represent habitats that usually favour disease development (Daamen et al 1989, Cook & Veseth 1991, Ou 1985). An exemplary case is the correlation of an increase in the severity of bacterial leaf blight with increased nitrogen input in rice (Reddy et al 1979). As a result, crop intensification is often perceived as inherently associated with enhanced pest problems. A more balanced perception is desirable; in rice for instance, the contrast between 'poor farmers' diseases' (e.g. brown spot, enhanced by water stress and/or limited nitrogen supply), as opposed to 'rich farmers' diseases' (e.g. blast), is well documented (Zadoks 1974, Ou 1985). It has been shown, however, that in wheat the magnitude of the detrimental effect of intensification practices on some pests is similar to that of the better-known beneficial effects of pesticides and resistant varieties (Daamen et al 1989).

Infrastructure and support services

There is controversy in plant protection circles over the degree of empowerment required for pest management decision making before a sustainable system is in place. One of the conclusions of a study commissioned by the Consultative Group on International Agricultural Research (CGIAR) was · that the introduction of seed-based technology (as exemplified by the green revolution) in the form of host plant resistance does not recognize the potential of farmers to address and solve their own pest problems (Teng et al 1985). For example, the Indonesian programme on rice IPM consists almost entirely of training farmers to give them the skills needed to understand and manage their environment (Wardhani 1992). The effort to train even 10% of all Indonesian farmers is a substantial one and there is always the risk of reversion to calendar-based spraying tactics. The system has not worked in other countries because there has not been the same level of infrastructural and resource support.

As technology impacts on cropping systems, the role of public sector plant protection services has to be re-defined. In Malaysia, an interesting development is the formation of 'rice estates', where smallholdings are amalgamated and managed using common practices over a relatively large contiguous area. Benefits are derived from the economics of scale, but the risk of pest outbreaks is also higher. The experience in the USA with increasing farm size has shown that farmers then delegate the monitoring of their crops to professionals owing to lack of time or perceived lack of expertise. In the USA, this also led to the growth of a substantial private crop protection sector in the 1980s. Asian countries in general have relatively well-developed plant protection systems in which the IPM concept appears well appreciated (Teng & Heong 1988). In other parts of the world, notably West and Central Africa, such services are particularly weak (Teng 1985). Even simple diagnostic laboratories appear not to be functioning in many African countries.

The relationship between the farmer and public sector crop protection services in making decisions concerning pest management offers an opportunity to develop sound approaches to plant protection. While farmers should be empowered with decision-making on their fields, crop protection services should address issues that involve communities, e.g. prediction of migration of brown plant hopper onto rice, and forecasting of first appearance of *Cercospora* leaf spot on sugar beet. These activities are relevant for a public service, and require skills and equipment not commonly available to farmers.

Farmer involvement in plant protection

Research on population dynamics, crop loss and pest management leads to recommendation of pest management practices. Many factors determine the implementation of practices at the farm level; a major one is the way farmers perceive pest problems. Pest management still relies heavily on non-durable components that affect the system, such as pesticides and non-durable resistance. Impermanent pest control components result in unsustainable pest management systems. One reason for the development and use of such components is their reassuring value from the farmers' stand point. A major determinant of the implementation of sustainable pest management practices is an adequate understanding of farmers' perceptions.

Fig. 4 represents a summary of a correspondence analysis performed on information gathered on farmers' perceptions during the integrated pest survey conducted by IRRI in Central Luzon, the Philippines (Elazegui et al 1990). The diagram shows a 'map' of farmers' perceptions as related to their background in terms of age, size of the farms and type of occupation of the land. It indicates that whorl maggot, rice bugs and defoliators are seen as much more of a problem by elderly farmers than by middle-aged (who are more concerned about stem borers) or young farmers. Tungro and brown plant hopper are concerns for all. Stem borer is a worry on small farms, whereas concerns about defoliators and whorl maggot are more often found on large farms. The diagram also suggests a difference between lease-holders, who have relatively few concerns except brown plant hopper and tungro, and owners and tenants, who appear to have more concerns. In summary, the diagram shows that depending on their background, farmers have different perceptions. When new

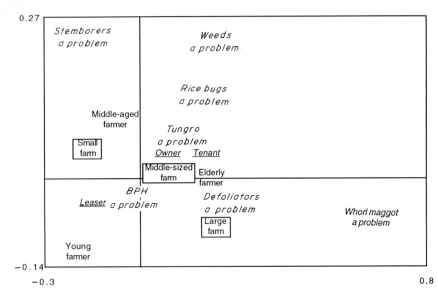

FIG. 4. A correspondence analysis of rice farmers' perceptions of their pest problems in Central Luzon, the Philippines. The sampled population consisted of 90 farmers. The graph represents their perceptions of pests using the two first axes generated by a correspondence analysis. Only some of the farmers' concerns about pests are shown here.

pest control practices are being introduced, farmers of different types should therefore be approached differently.

Economics of plant protection

Any sustainable form of plant protection must also be economically viable. Waibel & Engelhardt (1988) delineated a simple framework for evaluating methods of pest management. From a pay-off matrix including four combinations of recommendations and actions, criteria were developed that measure the economic consequences of decisions. Although this framework was developed for insect control by means of insecticides, it can be considered in the broader perspective of pest management by any tactical method.

Two recommendations may be considered (Table 3): to apply a control method (action, R) or not to apply it (no action, r), on the basis of an economic criterion (e.g. an economic threshold). The action has a cost, S. The farmer may decide to act (A) or not to act (a). The reduction in net return owing to a pest can be seen in two components: the first, P, can be avoided by the control action; the second, p, represents a residual (economic) effect of the pest. In the computations of the farmer's net returns, the losses due to the pest (P and/or p) and the costs of action (S) are subtracted from the net value of the crop, Y.

TABLE 3 A framework for evaluating pest management methods

	Decision	
Recommendation	Action (A)	No action (a)
Action recommended (R)	$AR = Y - S$	$aR = Y - (p + P)$
Action not recommended (r)	$Ar = Y - S - p$	$ar = Y - p$

Entries are economic returns. Modified from Waibel & Engelhardt (1988). See text for explanation.

This framework enables simple calculations to be made. For instance, the financial difference between the two scenarios, ar and aR, represents the economically avoidable loss, i.e. the gain that can be expected using the current control method. Similarly, the difference between Ar and AR is p, the losses that are not worth being controlled with this method (technically but not economically avoidable losses). The gap between AR and aR represents the optimum profit derived through this method, $(p + P) - S$.

The framework also allows consideration of future methods. AR represents the net returns of a farmer confronted with a problem and responding to it adequately. ar represents the net returns of a farmer who is no longer exposed to this problem. The difference $(ar - AR)$ therefore represents the gain that could be attained when introducing a new and more efficient control method.

A prerequisite for the calculation of such indices is an accurate measurement of inputs and outputs in all the considered pest management scenarios. This has to be carried out by means of on-farm trials. These trials should include the commonly used method and the improved one, together with plots on which no pest control is applied, in order to measure potential progress.

This type of analysis does not pertain to strategic decisions such as the deployment of resistance genes or the management of habitats for natural enemies of pests. All costs incurred by pest control methods cannot be accounted for using monetary units. As sustainability evolves as a new paradigm in the management of agricultural systems, new indicators are needed to monitor their performances (Wilken 1991), especially in crop protection.

The impact of plant protection practices on the environment and human health

Traditional cropping systems that do not have external inputs are often viewed in 'romantic' terms as being sustainable, even though they may not be because of their limited capacity to support human populations.

Several studies have examined the impact of pesticide use on the environment and human health (Pimentel 1983, Pimentel & Andow 1984, Pimentel et al 1991, Antle & Pingali 1992). These studies recognize that while the benefits of pest management to society today amount to several thousand million dollars each

TABLE 4   Estimated total environmental and social costs associated with pesticide use in the USA

Cause	Cost ($ million)
Human pesticide poisoning	250
Animal pesticide poisoning and contaminated livestock products	15
Reduction of natural enemies of pests	150
Resistance of pests to pesticide	150
Honeybee poisoning and reduced pollination	150
Losses of crops and trees	75
Losses of fish and wildlife	15
Government pesticide pollution regulations	150
Total	955

From Pimentel et al (1991).

year, there are also significant environmental and social costs associated with the use of pesticides. These include poisoning of humans and domestic animals, contamination of food and various impacts on agroecosystems and natural ecosystems. In the USA, the estimated environmental and social costs amount to $955 million each year (Pimentel et al 1991) (Table 4).

Concluding remarks

Ultimately, the relationship of plant protection systems that researchers recommend and that farmers adopt to sustainable development of the agricultural community should be recognized. Of particular concern are the relationships to social equity, to economic viability and to ecological stability. Plant protection systems have evolved relative to the broader political environment of government support. It is important that decision-makers in agriculture incorporate long-term ecological reasoning into their economic and policy formulation processes, especially in developing countries. Government policy in support of improved infrastructure (credit, subsidies) should be accompanied by actions that empower farmers, such as choice of non-standard crop varieties and practices. Green revolution technologies such as the use of pesticides and fertilizers have been too closely linked to public sector attempts at improving farming methods, often to the total exclusion of traditional practices. Even with rice, the world's most important food crop which feeds over 60% of the population and is grown in a relatively restricted geographical domain (about 90% in southern and South-East Asia), the hard question has to be asked whether new pest management technologies have resulted in any improvement in farming practices and in the living standards of the farmers. For sustainable development, we urge that plant protection should build on indigenous knowledge systems and involve farmers in the incremental changes to meet new objectives of production and productivity.

We have not discussed the role of energy requirements in different plant protection systems, in particular, exogenous, synthetic energy such as petroleum-based products. Studies on energy budgets (Cox & Atkins 1979, Spedding & Walsingham 1976) have shown that modern technologies are potentially less energy efficient than traditional ones, especially when the long-term objective of the system, i.e. its sustainability, is considered.

Our analysis of plant protection systems in terms of ecology, intensification, infrastructure development, communication, socio-economics and negative/positive externalities using these selected farming systems points out the potential of unsustainable farming systems evolving, if realistic plant protection is not taken into consideration.

References

Andow DA, Hidaka K 1989 Experimental natural history of sustainable agriculture: syndromes of production. Agric Ecosyst & Environ 27:447–462

Antle JM, Pingali PL 1992 Pesticides, farm health and productivity: a Philippine case study. Paper presented at the workshop on Measuring health and environment effects of pesticides, Bellagio, Italy, 30 March–3 April 1992

Cook RJ, Veseth RJ 1991 Wheat health management. American Phytopathological Society, St Paul, MN (Plant Health Manage Ser)

Cox GW, Atkins MD 1979 Agricultural ecology: an analysis of world food production systems. WH Freeman, New York

Daamen RA 1990 Pathosystem management of powdery mildew in winter wheat. PhD thesis, University of Wageningen, Wageningen, The Netherlands

Daamen RA, Wijnands FG, Van der Vliet G 1989 Epidemics of diseases and pests of winter wheat at different levels of agrochemical inputs. A study on the possibilities for designing an integrated cropping system. J Phytopathol 125:305–319

Elazegui FA, Soriano J, Bandong J et al 1990 Methodology used in the IRRI integrated pest survey. In: Crop loss assessment in rice. International Rice Research Institute, PO Box 933, 1099, Manila, the Philippines p 241–271

Glass EH, Thurston HD 1978 Traditional and modern crop protection in perspective. Bioscience 28:109–115

Mollison B 1988 Permaculture: a designer's manual. Tagari Publications, Tyalgam, New South Wales

Ou SH 1985 Rice diseases, 2nd edn. CAB International, Wallingford

Pimentel D 1983 Environmental aspects of pest management. In: Shemlit LW (ed) Chemistry and world food supplies: the new frontiers (Chemrawn II), proceedings of the international conference, Philippines, December 6–10 1982. Pergamon Press Reprints, Oxford, p 185–201

Pimentel D, Andow DA 1984 Pest management and pesticide impacts. Insect Sci Appl 5:141–149

Pimentel D, McLaughlin L, Zepp A et al 1991 Environmental and economic impacts of reducing US agricultural pesticide use. In: Pimentel D (ed) CRC handbook of pest management in agriculture, 2nd edn. CRC Press, Boca Raton, FL, vol 1:679-718

Reddy APK, Katyal JC, Rouse DI, MacKenzie DR 1979 Relationship between nitrogen fertilization, bacterial leaf blight severity, and yield of rice. Phytopathology 69:970-973

Rowe WD 1980 Risk assessment approaches and methods. In: Conran G (ed) Society, technology and risk assessment. Academic Press, New York, p 3-29

Savary S, Brissot F 1990 Epidemiology of foliar diseases and crop losses in groundnut in West Africa. Meded Fac Landbouwwet Rijksuniv Gent 55(2a):253-261

Savary S, Zadoks JC 1992 Analysis of crop loss in the multiple pathosystem: groundnut-rust-late leaf spot. III. Correspondence analyses. Crop Prot 11:229-239

Shane WW, Teng PS 1992 Impact of *Cercospora* leaf spot on root weight, sugar yield and purity of *Beta vulgaris*. Plant Dis 76:812-820

Shane WW, Teng PS, Lamey A, Cattanach A 1985 Management of *Cercospora* leafspot of sugarbeets: decision aids. North Dakota Farm Res 43:3-5

Spedding CRW, Walsingham JM 1976 The production and use of energy in agriculture. J Agric Econ 27:19-30

Teng PS 1985 A situation analysis of plant protection systems in West and Central Africa. Consultants report to the Plant Protection Service, FAO. FAO, Rome

Teng PS 1986 Action plan for the improvement of plant protection services in Central Africa. Final report submitted to the Plant Protection Service, FAO. FAO, Rome

Teng PS, Heong KL (eds) 1988 Pesticide management and integrated pest management in southeast Asia. Consortium for International Crop Protection, College Park, MD

Teng PS, MacKenzie DR, O'Laughlin J 1985 Impact of the international agricultural research centers on crop protection in developing countries. Consultant's report. World Bank/CGIAR (Consultative Group for International Agricultural Research)

Teng PS, Fernandez PG, Garcia R 1991 Indigenous plant protection practices in Philippine tribal systems. Report to GTZ Biological Plant Protection Project, Manila

Thalekar NS 1992 Integrated management of diamondback moth: a collaborative approach in southeast Asia. In: Ooi PAC et al (eds) Integrated pest management in the Asia-Pacific region. CAB International/Asian Development Bank, p 37-49

Thuy NN, Thieu Dv 1992 Status of integrated pest management programme in Vietnam. In: Ooi PAC et al (eds) Integrated pest management in the Asia-Pacific region. CAB International/Asian Development Bank, p 237-250

Waibel H, Engelhardt T 1988 Criteria for the economic evaluation of alternative pest management strategies in developing countries. FAO Plant Protection Bulletin 36:27-33

Wardhani MA 1992 Developments in IPM: the Indonesian case. In: Ooi PAC et al (eds) Integrated pest management in the Asia-Pacific region. CAB International/Asian Development Bank, p 27-36

Wilken GC 1991 Sustainable agriculture is the solution, but what is the problem? Board for International Food and Agricultural Development and Economic Cooperation (Occas Pap 14) Agency for International Development, Washington, DC

Zadoks JC 1974 The role of epidemiology in modern plant pathology. Phytopathology 64:918-923

Zadoks JC 1981 Crop loss today, profit tomorrow: an approach to quantifying production constraints and to measuring progress. In: Chiarappa E (ed) Crop loss assessment methods. CAB/FAO, Farnham, suppl 3:5-11

DISCUSSION

Neuenschwander: You highlighted the trade-off between production and sustainability. Ideally, we would like plants which have enormous yields and can be grown in a sustainable system, but in practice there is a trade-off. The international institutes, I'm talking for the International Institute of Tropical Agriculture (IITA), have for 18–20 years concentrated on the productivity component, probably rightly so. But now the perception is that we have come to a plateau, where production cannot easily be increased and sustainability becomes more important. Sociological surveys of farmers in West Africa have shown many times that the farmers' first concern is stability, not production. You may say sustainability is a romantic notion, but at least it is one favoured by the farmers.

Savary: Maybe we are a bit confused about terms. Africa represents a particular situation. When I spoke of romantic ways of viewing sustainability, I was primarily referring to situations like those present in many countries in South-East Asia. As far as West or Central Africa is concerned, stability is probably what matters first.

Neuenschwander: IITA finds that its high-yielding varieties of, for example, cassava are not always accepted by the farmers. They are being grown in certain areas: we publish that they are distributed over several hundred thousand farms or hundreds of thousands of square kilometres. But although those very good, high-yielding varieties are distributed over that area, the farmers have them as one variety among many. When you look at the contribution of these varieties to the overall production (values are very hard to get), the contribution is not as high as it could be.

Why? It is because of this trade-off. Those varieties do not just have higher yield, with everything else the same. Sometimes they do not taste the way the farmers want or they cannot be stored easily. Storage capacity is a part of sustainability. We are talking about very low yields in the farmers' fields and the farmers not having the means of reacting to disturbances. That's why IITA has stopped focusing on yield only. Yield is still important, but for a variety to be successful, and that means adoption by the farmers, we have to look at the other components. In this context we find that the farmers are very much looking for 'sustainability'.

Zadoks: ICRISAT had the same experience in West Africa. It developed beautiful varieties of maize and sorghum but they were not accepted.

Varma: Under West African conditions, when people grow crops for themselves, they are definitely more concerned about the taste and the storage quality. But for crops that they grow for export this is not so important. For example in northern Nigeria, a single variety of tomato, called Roma, was introduced, accepted and is being cultivated on a very large scale. In India, high-yielding dwarf varieties of wheat were accepted so enthusiastically by farmers that the area under these varieties increased from 4 ha in 1964–1965 to four million hectares in 1971–1972 (Swaminathan 1978).

Savary: The analysis I showed of the farmers' perception of pests and pest problems (Fig. 4) would be impossible in West Africa, for the simple reason that in many cases farmers do not perceive diseases as problems. Leaf spot, for example, was introduced to Africa simultaneously with groundnut. The farmers have been confronted with this disease since the crop has been grown in Africa. When trials were started with fungicides in farmers' fields, the farmers became very angry, because what they saw was not groundnut as they knew it. It was a crop that had green leaves at the end of the cropping season and therefore could not be harvested. In their experience, groundnut should be harvested when the crop has no leaves left.

Varma: One reason for farmers' lack of interest in crop protection measures is that generally these are not cures, they are prophylactic methods. So whatever treatment you apply, you cannot show the kill to the farmers as entomologists are able to show them. Therefore the farmers are not so keen to use control measures. Extension workers have to play a very important role in implementing disease management practices.

Jayaraj: Farmers' perceptions vary with the innovations or the technologies. Improved crop varieties are readily perceived and adopted. With nutrient management, again perception and acceptance occur rapidly. But plant protection, particularly for grain farmers, is too complicated and too specialized for easy perception. If the farmer is to be enabled to take decisions on pest management, he needs counselling and guidance from extension workers as well as from consultancy agencies. That's why we often find considerable differences between the farmers' perceptions of a particular plant protection technology and their perceptions of crop varieties or crop improvement technology.

Another situation is where the farmer is concerned about a particular pest or disease. When he tries to solve that with the help of the consultancy or extension agency concerned, he is not aware of what will happen next. He is not able to perceive the sequence of pest and disease problems that will arise in the crop, in neighbouring crops or in the succeeding crop. The farmers should be encouraged to consider IPM-related problems for a whole cropping system. Community efforts will also have to be brought in. IPM is field specific: we may have to organize the farmers through community approaches to manage the situation over a larger area. Unless these sociological aspects are considered, we will not be able to make progress in the adoption of the modern technology in plant protection.

Norton: John Stonehouse looked at beans in Columbia and how farmers made decisions about disease control. He found they had actually developed their own thresholds; these hadn't come from scientists, the farmers had done it themselves. His hypothesis was that because there hadn't been a change in varieties (they were using traditional varieties with which they had 20–30 years experience), their perceptions were fairly good, they didn't panic when they saw disease, and they waited until it reached a threshold before they did anything about it. One might contrast this with the situation in rice where there is constant change and therefore farmers' perceptions aren't so good and perhaps the farmers are more risk averse.

Savary: Farmers do have their own thresholds. However, because of changes in cropping practices and pest patterns, the threshold theory in plant protection is to some extent at stake (Zadoks 1985). One example is groundnut foliar diseases in West Africa, where there is a wide range of cropping practices and a series of pathogens, one recently introduced to the region (Savary & Zadoks 1992).

Norton: When you discuss the farmers' perceptions of problems, to what extent is it perception of a problem and to what extent is it their attitude to risk? Obviously, larger farmers are less likely to be risk averse, if they have higher incomes and so on, than smaller farmers. (That may not be the case; farmers with large farms may be deeper in debt. I don't know.) Attitude to risk and perceptions would interact in a variety of ways. Do you have any information on this?

Savary: The survey tried to determine what were the farmers' main worries. Pests were one component; there were others. This was addressed in unconventional, individual discussions with 90 farmers. In the analysis I showed, the responses were an indication of yes or no; for example, whether tungro had been mentioned by farmers or not (Fig. 4). Tungro was in the centre of the diagram, indicating that tungro was a common worry among all the farmers in the survey.

Jones: You classified farmers as small farmers, large farmers, or according to the way they held the land (Fig. 4). They were seen to have different perceptions as to what their main problem was. Was this because they had different problems, or because they all had the same problems but they assessed the risks differently? If they had different problems according to the size of their farm, then that's important. If they had different perceptions, even though their problems were all the same, that's important too. To distinguish between these two cases is also important.

Savary: This correspondence analysis is given as a map; only the farmers' worries are shown, not their actual problems. In fact, when the survey was conducted there was no tungro, nor was there any problem with brown plant hopper. But tungro and brown plant hopper were still mentioned as a major problem by a large proportion of the farmers.

Jones: Is there any association between these different types of farmers and the problems that occur in their fields, or is it all in their heads?

Zadoks: It is in their heads.

Upton: The assessment of farmers' perceptions and attitudes and objectives is very difficult. I don't really know how you can say that the farmers are not concerned with maximizing income but are concerned with sustainability in West Africa. We have a Nigerian girl, Ovuevuraye Dicta Akatugba, writing her PhD on farmers' objectives. She has used quite sophisticated techniques to investigate farmers' objectives, and some interesting things have come out. The primary objective stated, almost invariably, is to produce food for the family. Profit comes somewhere below that. Investigation of attitudes to sustainability or to risk is very difficult. We can't even agree about what we mean by sustainability, so how can we get a farmer to explain to us what he understands by sustainability? Similarly with attitudes to risk: it's taken quite a sophisticated study by Hans Binswanger (1980) to show that Indian farmers are risk averse.

In addition, we are prone to interpreting the answers farmers give in the way we wish. For instance, the expressed desire to produce food for the family can be seen simply as a productive attitude—the aim of producing enough income and enough produce to feed the family—or it can be seen as a risk averse attitude—the aim of producing enough food even in a fluctuating situation. So there are problems of interpretation. Overall, this whole business of farmers' perceptions and objectives is a very difficult area to investigate.

Jones: I'm a little uneasy that risk aversion is being equated with stability and stability is being equated with sustainability. I would try to distinguish between these things.

Norton: There's a lot we need to do in terms of understanding how farmers make their decisions. I'm not sure whether the detailed risk analysis that Martin Upton was talking about is necessarily the best way. Paul Richards in the UK talks about the performance of farmers. Each season, it's a different performance: the farmer has a whole series of techniques he can use, and he varies them depending upon the season.

Serge Savary mentioned the mixed cropping systems with maize in West Africa. In Kenya, we found farmers whose preference is to grow single crop maize using high fertilizers, and to weed it properly and get a high yield. If for some reason in the season, through illness or whatever, they find that they have to plant a late crop that they are not going to be able to weed properly, then they will intercrop. They plant beans in with the maize and leave it alone. So the performance each season depends upon their particular circumstances. We have to understand this a lot better before we can undertake appropriate research.

Martin Upton is being a bit critical of us in the sense that what we have to do is just get a better feel for the situation. We don't need to know the degree of risk attitude, we just need some general impressions. General impressions are better than nothing. Some entomologists (I'm sure nobody here!) would work out thresholds on the experimental station, then transfer those to the farmer. We are saying we need to understand the farmer a little better, for example to understand his general risk attitude. In Vietnam, many of the farmers were previously in the state system; they were virtually technicians, they did as they were told. Now, suddenly, they are decision-makers and they are extremely risk averse. It may be perceptions, it may be risks. One needs to find out what the problem is before one can take appropriate measures.

Rabbinge: Farmers' attitudes to risk aversion can be determined by different types of investigation. The other side of the coin is how you are improving the quality of their decision making by making explicit what risks pertain in particular situations. In different crop protection systems, we are trying to make explicit what risk there is in particular situations, depending on the expected yield, on the crop development stage, on the presence of biological control agents, on the prevailing weather and things like that. One can calculate in advance the risk of what yield depression one may expect. That gives the farmers the choice of accepting that risk or paying a premium by implementing a preventative agronomic measure.

Waibel: We have tried to apply the pay-off matrix to the field of crop loss assessment. This is one way to derive something concrete from this confusing set of definitions of crop loss.

In your experiments, you compared actual yield and attainable yield. Recently, I have talked to Dr Pingali, an economist at IRRI, who is publishing a book on the economics of pest management in rice. He has compared results achieved by farmers practising IPM or natural control or using prophylactic treatments (which can mean as many as 17 applications of insecticide in rice). He found no significant difference between farmers who don't spray and those using IPM. However, there is a significant increase in yield in fields treated with prophylactic measures compared with those managed using the other strategies. So I wonder whether the concept of attainable yield is really useful. Where you place this attainable yield is rather arbitrary. 17 applications of insecticide are definitely not economical or ecologically desirable.

Savary: Professor Zadoks' definition of the attainable yield was the one used in this paper, for groundnut as well as for rice.

Zadoks: What we envisaged as an attainable yield is not exactly the same as what the FAO envisaged. FAO envisaged attainable yield as the yield one could get in experimental plots under the best possible husbandry conditions given the limitations of the place. In my view, attainable yield is what a good farmer can get under relatively favourable conditions, within the constraints of his particular situation. My definition is oriented toward the farmer, not the experimental situation. I admit it's still arbitrary, but the difference between experimental field plots and farmers fields can be 1 ton ha^{-1}. Usually, yields from experimental plots are 1 ton ha^{-1} greater than those from farmers' fields; the converse also exists. In The Netherlands, yields of 10 tons of wheat ha^{-1} without any treatment are achieved, which is much more than you get on any experimental plot.

Savary: The operational definition for the rice crop loss database is that if you consider one pest at a time, you have basically two types of treatments—the plots where the pest has been introduced and the plots where the pest has not been introduced. The experiments have been conducted in the farmers' fields, as well as in experimental fields.

In the process of analysing this crop loss database, we are considering using categorized information—in other words, classes of yield levels. For a long time the idea of yield response surface has been used to explore yield variation as a result of many factors, including pests and diseases. However, we are including 'discontinuous' components like water management, direct seeding, fertilizer use and the potential yield of varieties. Therefore we need to handle this in a categorical way.

Rabbinge: When we talk about attainable and actual yields, we have to distinguish three different things: the production situation, the production level and the production methods. The production situation is dictated by the land reclamation level—the quality of the soil. The production situation determines the yield you can achieve given the limitations of the particular location. I explained yesterday how you can simulate or compute expected potential yields (Rabbinge, this volume).

The production level is the level you are trying to achieve. Even in a poor production situation, say in a sandy soil where there is low water availability during the growing season, you still can have a high production level by having a lot of irrigation, by having a lot of external inputs. But the risk is that you overuse your external inputs. The production level should be tailored to the specific circumstances of the production situation.

The third point is the production method. How will you apply the treatments? What crop rotation are you using? How are you seeding? What is the seed density? There we should always try to aim at the best technical means, i.e. inputs are used in a way that, per unit of product, each input is minimized such that the efficiency of use of all the other inputs is maximized. This depends again on the production level and the production situation.

I would also like to make a point concerning intensification. If by intensification you mean that you are trying to have high productivity per unit area, I completely agree. If you mean intensification per unit of product, then I completely disagree. I think you mean per unit of area; that can mean that per unit of product you are decreasing the inputs. This is important because in much of the industrialized world these are taken to be synonymous, and they are not. We should aim at reducing some inputs per unit area, while improving the efficiency of the inputs per unit of product; and that's normally at good production situations at high production levels. Then there is the problem of stability. If the rise in productivity occurs too soon or too rapidly and not steadily, then you have instabilities. This was seen in the former Soviet Union. The rise in productivity was stimulated by all types of external inputs, such as nitrogen fertilizers and new varieties of plants. But the stability of the yields was falling, because the knowledge and the technological insights were not sufficient to combine the productivity rise with the proper way of farming in such a way that you have sufficient stability.

Savary: I agree completely.

Varma: The size of fields in relation to the susceptibility of the crops to diseases is a debatable point. Large fields generally have less disease; it depends on the source of infection. If it comes from outside, in a larger field, there is less disease. However, if there is a single genotype in a very large area, then you run a risk of sudden strain development of a pest which may wipe out the crop from the area.

Zadoks: Mycologists are convinced that large fields are more risky than small fields; virologists often maintain the converse. There must be some technical reasons behind it which I have not yet found, but part of it is the transmission efficiency by insects.

Varma: Van der Plank observed that maize streak virus destroyed the whole crop in small fields but not in large fields in South Africa (Broadbent 1964). I found a similar situation for tomato leaf curl virus in West Africa (Varma 1984). There are several reasons for a lower incidence of viral diseases in large fields; one is transmission efficiency, another is the source(s) of infection.

References

Binswanger H 1980 Attitudes towards risk: experimental measurement in rural India. Am J Agric Econ 62:395–407
Broadbent L 1964 Control of plant virus diseases. In: Corbett MK, Sisler HD (eds) Plant virology. University of Florida Press, Gainesville, FL, p 330–364
Rabbinge R 1993 The ecological background of food production. In: Crop protection and sustainable agriculture. Wiley, Chichester (Ciba Found Symp 177) p 2–29
Savary S, Zadoks JC 1992 Analysis of crop loss in the multiple pathosystem groundnut-rust-late leaf spot. III. Correspondence analysis. Crop Prot 11:229–239
Swaminathan MS 1978 Wheat revolution—the next phase. Indian Farming 27(11):7–11
Varma A 1984 Virus diseases of fruit and vegetable crops in Nigeria. FAO, Rome (FAO Tech Rep 10 AG:DP/NIR/72/007)
Zadoks JC 1985 On the conceptual basis of crop loss assessment: the threshold theory. Annu Rev Phytopathol 23:455–473

Integrated management of plant viral diseases

Anupam Varma

Advanced Centre for Plant Virology, Division of Mycology and Plant Pathology, Indian Agricultural Research Institute, New Delhi 110 012, India

Abstract. Viral diseases of plants cause enormous economic losses particularly in the tropics and semitropics which provide ideal conditions for the perpetuation of viruses and their vectors. Intensive agricultural practices necessitated by the ever-increasing demands of the rapidly growing population and the introduction of new genotypes, cropping patterns and crops have further aggravated the problem of viral diseases. Many diverse approaches have been tried to minimize the losses caused by these diseases. The approaches are mainly based on avoidance of sources of infection; avoidance or control of vectors; modification of cultural practices; use of resistant varieties obtained though conventional breeding procedures; cross protection; systemic acquired resistance; and use of transgenic plants containing alien genes that impart resistance to viruses. Although the use of resistant varieties has been found to be the most economical and practical, for effective management of viral diseases an integrated approach is essential in sustainable agriculture. Development of integrated management practices also requires correct identification of the causative viruses, because symptoms can be misleading, and adequate understanding of the ecology of viruses and their vectors.

1993 Crop protection and sustainable agriculture. Wiley, Chichester (Ciba Foundation Symposium 177) p 140–157

Diseases of plants caused by viruses and virus-like organisms result in enormous economic losses (Varma 1976). The losses are greater in the tropics and semitropics, which provide ideal conditions for the perpetuation of viruses and their vectors. For example, mung bean yellow mosaic geminivirus is estimated to cause an annual loss of US $300 million in the production of mung bean, black gram and soybean (Varma et al 1992a) and cassava mosaic geminivirus causes economic losses equivalent to US $2000 million annually in Africa alone (Fauquet & Fargette 1990). Losses also occur in the temperate regions (Hull & Davies 1992). The losses are even more alarming in perennial plants, where there is a steady decline as a result of infections by viruses and virus-like organisms (mycoplasma-like organisms, bacteria-like organisms, viroids, etc). During the last fifty years such diseases have resulted in the destruction of over 200 million citrus, cocoa, coconut and sandal trees in various parts of the world.

Intensive agricultural practices, necessitated by the ever-increasing demands of a rapidly growing population, have further aggravated problems of diseases and pests (Varma & Verma 1991). The increase in the spatial distribution and intensity of diseases caused by tospoviruses and whitefly-transmitted geminiviruses are just reminders of the enormity of the problems and resultant instability in agricultural production. Therefore, for sustainable agriculture, appropriate management of viral diseases is essential.

Four distinct approaches to agricultural sustainability have been suggested, but for countries like India the only option is to improve the availability of food *per capita* to an adequate level through increased production (Varma & Sinha 1992). For this objective and sustained agricultural growth to be achieved, all sorts of efforts will have to be made, from judicious use of technological developments to minimization of the losses caused by biotic and abiotic stresses.

Control of viral diseases is more difficult than of those caused by other pathogens, because of the complex disease cycle (Fig. 1), efficient transmission and lack of viricides. The effective management of viral diseases requires integration of management practices, such as avoidance of sources of infection, control of vectors, modification of cultural practices and resistance of host plants.

Avoidance of sources of infections

The main sources of viral infection are infected planting material, collateral hosts, adjoining crops, volunteers and ground keepers. Occasionally, vectors are also a source of primary infection.

Use of virus-free seed

Over 120 viruses are known to be transmitted through true seeds to a varying extent. Seeds carrying viruses not only result in a poor crop stand but also

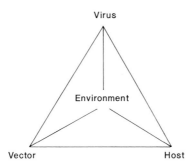

FIG. 1.   Viral disease triangle.

provide inoculum for secondary spread of the virus. The extent of secondary spread varies from virus to virus, depending on the efficiency of transmission by the vector(s). In the case of lettuce mosaic virus, 1.6% infection through lettuce seed resulted in 30% final incidence, whereas for cowpea (vein) banding mosaic virus 4.5% seed transmission was required for a similar final incidence (Varma 1976).

The viruses are carried through seeds either as surface contaminants, for example tomato mosaic virus (ToMV) in tomato, or in testa, as is southern bean mosaic virus (SBMV) in cowpea and beans, or in the embryo, as in black gram mottle virus (BMoV). The mere presence in seed does not ensure transmission. For example, BMoV is transmitted through seed only when the embryonic axis of black gram (*Vigna mungo*) seed has a minimum of 60 ng virus (Varma et al 1992b), whereas SBMV is not transmitted through the seeds of cowpea var V 152 even when present in very large amounts in the testa.

Not much attention has been given to curing virus-infected seeds. Treating seeds with dry heat, chemicals (e.g. malic-hydrazide, 2-thiouracil) or γ-irradiation, or storage for various periods has helped to cure seeds of different viral infections (Varma 1976) but the large-scale application of such techniques is difficult. The most practical approach for avoiding this source of infection is to produce seed in virus-free areas and provide certified seeds to the growers, as has been recommended for pea seed-borne mosaic virus (Varma et al 1991a).

Use of virus-free vegetative propagules

Primary infection through the planting material is even more serious in vegetatively propagated crops like potato, cassava, sweet potato, sugarcane, citrus and banana. Viral diseases in all these crops cause considerable losses in economic yields annually. For vegetatively propagated plants, the use of healthy (virus-free) planting material is imperative. It requires careful testing, because even propagules obtained from healthy looking plants may not be free from infection, as was found for cassava infected with Indian cassava mosaic virus (Malathi et al 1989). Determination of freedom from infection by a pathogen is restricted by detection limits. Recent biotechnological methods, like the immunocapture polymerase chain reaction which is 5000-fold more sensitive than ELISA (enzyme-linked immunosorbent assay) (Wetzel et al 1992), have made it possible to detect even a single pathogen in propagating material. Such sensitive diagnostic techniques will be very useful in developing programmes for the use of healthy planting materials.

Heat treatment helps to cure planting materials of viruses like potato S, potato leaf roll and sugarcane mosaic, because the viruses are inactivated at temperatures not lethal to plants (Varma 1976), but large-scale application of heat therapy has been possible only to a limited extent, owing to practical difficulties in treating planting material in bulk. The best approach is to produce virus-free planting material. Many useful programmes have been developed for producing healthy potato seed in areas or seasons which do not favour the accumulation and movement of the vector population.

In India, disease-free potato seed production is also undertaken on the plains of northern India, which have a milder climate, by taking advantage of the smaller population of aphids between September and January. The fields are sown in late September and harvested not later than the first week of January. The aphid population usually starts to build up and migrate during the second half of January (Fig. 2). This technique, commonly referred to as the 'seed plot technique' has been a grand success in increasing potato production in the country.

These strategies are not possible in long-duration crops like cassava, banana and citrus. For such crops, it is essential to have special programmes for eradicating diseased plants and other sources of infection and replanting with healthy materials (Supriyanto & Whittle 1991).

In vegetatively propagated crops, infection of an entire clone/cultivar is common. Virus-free plants have been obtained by meristem culture to evaluate their performance. Success in eliminating a viral infection depends on the size of meristematic tissue used for culture (Kassanis & Varma 1967). This technique has been widely used to obtain virus-free clones of crops like cassava, banana and sugarcane. In citrus, *in vitro* 'shoot-tip grafting' has been very successful in producing virus-free plants (Vogel et al 1988). Where meristem or shoot tip culture does not work, a combination of heat or chemotherapy with tissue culture has been found useful.

Avoidance of collateral hosts and volunteers

Collateral hosts (weeds and related crop plants) play an important role in the perpetuation of viruses. For viruses transmitted through nematodes, weeds help not only in perpetuation but also in spatial distribution, because these viruses are also usually transmitted through the seeds of their host plants. Control of weeds, therefore, reduces the losses caused through competition and minimizes the losses due to viruses. Weeds are very important reservoirs of viruses belonging to groups like cucumo-, poty- and geminiviruses. Removal of malvaceous weeds (e.g. *Malva sylvestris* and *Sida rhombifolia*) around plots of bhindi (*Abelmoschus esculentus*) and observing a closed period before sowing the principal crop reduced the incidence of damaging bhindi yellow vein mosaic virus (Varma 1976).

Volunteer plants and overlapping cropping sequences also help maintain continuity of the viral life cycle. In North America, wheat streak mosaic virus (WSMV) is a serious problem. The causal virus and its mite vector *Aceria tulipae* overwinter on self-sown volunteers and move to wheat in the spring. Late sowing of the winter crop was suggested as a possible method

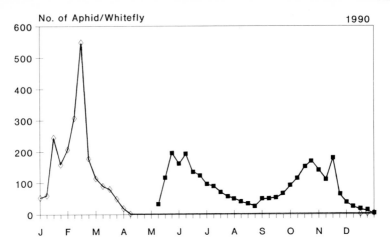

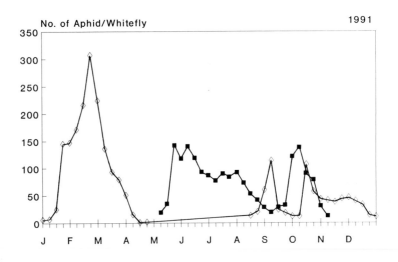

FIG. 2. Population of aphids and whitefly under Delhi conditions in 1990 and 1991 (K. P. Srivastava & A. Varma 1992, unpublished data). ◇, aphids; ■, whiteflies.

for control, but this reduces economic yields because the plants in the spring are more tender and susceptible to WSMV than those sown at the normal time. The best way to minimize the losses caused by this virus is through the removal of volunteer plants (Hunger et al 1992).

  Plants infected early in a crop provide a source of infection for secondary spread of viruses. Rogueing of such plants has been useful for managing bunchy top of banana in Honduras and swollen shoot of cocoa in West Africa. Rogueing in annual crops is not of much importance, particularly if the virus is transmitted efficiently.

Avoidance and control of vectors

Avoidance of vectors

This is a good approach, as discussed earlier for the production of healthy seed potato in northern India, but it cannot be used widely because the periods in which vectors can be avoided are usually short. In North India, mung bean yellow mosaic virus (MYMV) is a major constraint in the production of mung bean during *kharif* (wet season), which is the main growing season for this crop. Short duration varieties of mung bean were developed, which could be fruitfully cultivated during the *zaid* (summer season) under irrigation so that losses due to MYMV were avoided. However, in recent years, as a result of changes in the cropping pattern, the vector has started appearing early and causing a heavy incidence of MYMV even in the *zaid* crop, resulting in a decline in the area under summer mung (Varma et al 1992a).

Chemical control of vectors

Chemical control of the vectors is the most commonly used approach for the management of viral diseases. It has been tried either at the source of the vectors to prevent primary spread of the virus and/or in the crop to minimize both primary and secondary spread. Chemical control of vectors in the crop is particularly effective for viruses transmitted by vectors in a persistent manner, which require longer acquisition and transmission access periods. For viruses that can be acquired and transmitted within minutes, chemical control is less effective, but it does reduce the overall incidence of the disease by checking the build-up of the vector population in the crop.

For chemical control, timely application is very important. In the UK, sugar beet yellows disease is managed very effectively by applying insecticides whenever a warning is issued on the basis of the vector population (Varma 1976). The success of such control measures depends on efficient monitoring and information technology.

Whitefly-transmitted geminiviruses are more difficult to control by chemical treatments owing to the agility of the vector and the relatively short acquisition and transmission access periods required, although these are persistently transmitted viruses. The best disease control and improvement in economic yields have been obtained by treatments with systemic insecticides (Fig. 3; Sharma & Varma 1982a, Varma et al 1992a).

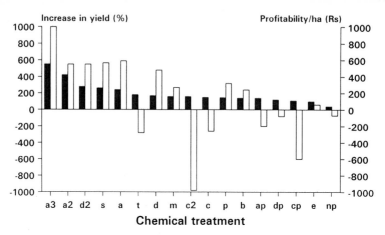

FIG. 3.   Economics of various chemical treatments in production of black gram cv. S1.
a, a2, a3 = 1, 2, 3 applications of Aldicarb; ap, Aldicarb + Paraffin oil; b, Diazinon;
c, c2 = 1, 2 applications of Carbofuran; cp, Carbofuran + Paraffin oil; d, d2 = 1,
2 applications of Disulfoton; dp, Disulfoton + Paraffin oil; e, Ekalux; m, Meta-
systox + Paraffin oil; np, Neem cake + Paraffin oil; p, Paraffin oil; s, Sunspray oil. Dark
columns, increase in yield; light columns, profitability (H. Subrahmanyam & A. Varma
1979, unpublished results).

Biological control of vectors

Biological control of vectors has not received the attention it deserves. It was
very successful in checking the spread of carrot mottley dwarf virus in Australia
and New Zealand through the introduction of the wasp *Aphidius*, which reduced
the population of the vector *Cavariella aegopodi*. It, however, does not work
in Europe (Varma 1976). Similarly, the spread of citrus greening disease, caused
by a bacteria-like organism, was successfully restricted through the biological
control of its psyllid vectors *Diaphorina citri* and *Trioza erytreae* in the Reunion
Islands. Within 24 months of the release of 30–50 eulophid ectoparasites
Tetrastichus dryi and *T. radiatus* per km² the population of psyllid vectors was
reduced and spread of greening disease checked. *T. radiatus* could not control
the population of *D. citri* in India, owing to the presence of several secondary
parasites that were not present in the Reunion Islands (Aubert & Quilici 1984).

Prevention of viral transmission by vectors

Spraying plants with mineral or vegetable oils reduces infection by the virus
transmitted by aphids in a non-persistent manner (Fig. 4) and also by those
transmitted by whiteflies. It is not, however, very effective against the persistent
type of aphid-borne viruses (Sharma & Varma 1982b). It is a useful approach,
particularly for vegetable crops, as it avoids the use of toxic chemicals.

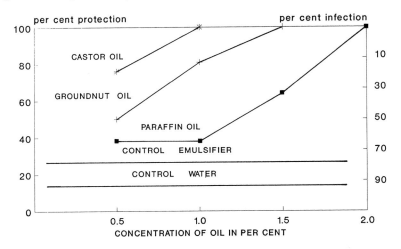

FIG. 4.   Effect of oil on transmission of water melon mosaic virus by aphids to vegetable marrow (Raychaudhuri & Varma 1983).

Modification of cultural practices

Suitable modification of cultural practices can help greatly in minimizing the incidence of viruses, particularly those with a narrow host range. Sterility mosaic 'virus', which is transmitted by an eriophid mite, *Aceria cajani*, causes serious economic losses in India. The host range of both the virus and its vector is restricted to pigeonpea, but they perpetuate on volunteer plants and ratoon/overlapping crops. Removal of volunteers and prohibiting ratoon/overlapping crops, even for one season, would break the disease cycle and prevent the losses caused by sterility mosaic (Varma 1976).

Keeping fields free from weeds is very useful, especially for viruses which have a wide host range or are also transmitted through the seeds of weed plants. Leaf curl caused by a whitefly-transmitted geminivirus is the most important disease of tomato in the tropics and semitropics. In Nigeria, nearly 100% of plants were infected with leaf curl in farmers' fields in the humid tropical parts of the country where control of secondary hosts is extremely difficult. In the arid northern regions, the incidence was less than 2.5% in farms which were kept completely free from weeds, compared with 90% in smaller fields surrounded by weeds (Varma 1984).

Various types of mulches have been used to reduce the incidence of viral diseases. Polythene mulches protect plants from aphid-transmitted viruses by repulsing aphids owing to reflection of heat. Mulches also help the growth of plants by conserving moisture, increasing soil temperature and eliminating competition by weeds (Vani et al 1989).

Intercropping of legumes with cereals of similar height protects crops from aphid-borne viruses. Soybean mosaic virus-induced seed mottling was 4% in soybean monoculture compared with 2% in mixed crops with dwarf sorghum cv G522 DR (Bottenberg & Irwin 1992). The incidence of cowpea (vein) banding mosaic virus in cowpea was 12.6% in monoculture but only 3.0% in mixed crops with maize (Sharma & Varma 1984).

Use of resistant varieties

This is the most important and practical approach to managing viral diseases. Genetic resistance to plant viruses may operate through resistance to vectors, transmission by vectors/seeds, virus multiplication, symptom development, intercellular movement, hypersensitivity or immunity to virus infection. Immunity is the most desirable form of resistance for managing viral diseases, because no detectable multiplication of the virus occurs in such plants. Two contrasting 'positive' and 'negative' genetic models have been suggested to explain resistance to viruses in plants (Fraser 1987). The positive model postulates that the resistant plants contain a gene or genes which prevent viral multiplication. The negative model proposes that such plants lack a gene or genes required for susceptibility. Although there is no conclusive evidence in support of either model, the non-specificity of immunity and the specificity of susceptibility suggest the negative model. This is also supported by the occurrence of viral diseases like cowpea golden mosaic, cotton leaf curl and okra leaf curl in India, which was earlier free from such diseases although the causal viruses existed in low numbers in the wild, owing to the inadvertent introduction of 'susceptibility' genes. Both the positive and negative models, however, may determine various other forms of resistance (Varma et al 1991b).

Most (approximately 80%) 'varietal' resistance to viruses is genetically simple, with resistance and susceptibility segregating in a typical Mendelian system involving a single locus (Khetarpal et al 1990). In some cases, more complex interactions with two or more loci have been detected, but given the few host–virus interactions for which the genetics of resistance has been studied and the limited parameters (Rajamony et al 1990) used to determine resistance, any generalization about the mechanism of resistance to viruses is difficult.

Resistance genes have been identified for nearly 130 host–virus combinations, but there is no set pattern for the mechanism of resistance. Of the 57 host varieties with monogenic resistance to one or more of 29 different potyviruses, 47% are dominant, 44% recessive and the remainder incompletely dominant (gene dosage dependent). Similarly, of the three different hosts identified to have monogenic resistance to cucumber mosaic virus, two are dominant and one is recessive. Thus, the mechanism of resistance varies not only from virus to virus, but even for the same virus in different hosts. For a long time, resistance to viruses was considered to be more durable than that to the other groups of pathogens, but in recent years resistance-breaking strains (isolates) of viruses have been reported for over 75% of the resistant host–virus combinations tested. The frequency of resistance-breaking strains is greater for dominant than recessive or incompletely dominant genes. The rapid development of resistance-breaking strains and, in many crops, the lack of sources of resistance makes breeding for resistance to viruses a difficult task (Varma et al 1991b).

Resistance may break down through the development of new biotypes of vectors or new strains of viruses. Many high-yielding cultivars of rice resistant to rice tungro virus (RTV) are widely grown in the Philippines. Resistance to RTV in some of these cultivars has been overcome by the vector leaf hopper (*Nephotettix virescence*) becoming able to feed on and colonize these cultivars continuously for several years (Dahal et al 1990). New strains arise frequently in viruses with divided genomes, like whitefly-transmitted geminiviruses. It is estimated that resistance to mung bean yellow mosaic virus in mung bean breaks down in less than six years (Varma et al 1992a), perhaps because of the occurrence of natural recombinants.

In many crops, sources of resistance to viruses are available within the species. For many others they are found in wild relatives, particularly at the centres of origin, owing to co-evolution of the crops and their pathogens. Useful resistance in crops has also been found in secondary centres of diversity (Lenne & Wood 1991). Introduction of new crops and extensive movement of germplasm, both essential for agricultural development, bring in pathogens and diseases new to either the crops or the region; cocoa swollen shoot is an example of the former situation (Thresh 1981) and cowpea golden mosaic in India (Varma et al 1992a) of the latter. In neither case did the virus move with the germplasm. For sustained agricultural development, careful efforts will be required to identify and use resistance genes available in the wild germplasm.

In crops where resistance is not available in desirable cultivars, the identification of rate-reducing resistance would be helpful.

Cross protection

Although the mechanism of cross protection—the ability of a mild strain to prevent infection by a severe strain—is not clearly understood (Gibbs & Skotnicki 1986), it has been successfully used in different countries for protecting citrus, tomato and papaya from infection by severe strains of citrus tristeza, tomato mosaic and papaya ringspot viruses, respectively. The most extensive application of cross protection has been in the citrus industry. Such protection, however, is not useful for sustained growth as (a) the plants may not be protected against all the severe strains, and (b) the protected plants may be predisposed to other viruses/pathogens. In Florida mild isolates of citrus trizteza virus failed to protect sweet orange on sour orange rootstock from infection by severe isolates, leading to decline. In South India, protected lime trees yielded less than the unprotected trees (S. R. Sharma 1988, unpublished findings). Satellite-containing mild strains of cucumber mosaic virus provided protection from the severe strain by 22 to 83% and increased yields by up to 56% in peppers in China (Tien et al 1987). It is, however, risky to introduce a large population of a virus that is efficiently transmitted and has a large host range into an ecosystem.

Systemic acquired resistance

Many compounds, such as plant growth hormones, nucleic acid base analogues, microbial metabolites, alkaloids, drugs and plant extracts have been tested for the ability to inhibit infection by viruses (Varma 1976). Treated plants develop non-specific systemic acquired resistance (SAR), which is associated with systemic induction of pathogenesis-related (PR) proteins, like chitinases and glucanases. Similar resistance also develops in the hypersensitive response and to damage caused by feeding insects. Recent evidence suggests that salicylic acid, systemin and ethylene act as endogenous molecular signals in SAR (Enyedi et al 1992). Endogenous salicylic acid increases more than 20-fold in leaves of tobacco (*Nicotiana tabacum* cv. Xanthi) inoculated with tobacco mosaic virus and fivefold in uninoculated leaves. Practical application of SAR for controlling viral diseases, however, requires further understanding of the mechanism at the molecular and cellular levels.

Resistant transgenic plants

Biotechnological developments have opened up the possibility of developing plants resistant to specific viruses through transfer of alien genes, particularly of viral origin, using a variety of systems. Several strategies, like transferring DNA copies of satellite RNAs that reduce symptom expression, non-structural gene sequences, defective interfering molecules, ribozymes or antisense RNA, have been tried (Hull & Davies 1992). The most effective approach has been coat protein-mediated resistance (Varma & Sinha 1992). This has been efficient against a large number of viruses, including those belonging to important groups like cucumo-, potex-, poty-, tobamo- and tobraviruses.

The mechanism of resistance is not well understood. It appears to vary in different host–virus combinations. In the case of cucumber mosaic virus, the transgenic plants that express larger amounts of viral coat protein are more resistant than those expressing a smaller amount (Quemada et al 1991), whereas the reverse was found for some potyviruses (Lindbo & Dougherty 1992). Coat protein-mediated resistance can be developed for more than one virus simultaneously. For example, potato cv Russet Burbank transformed with coat protein genes for potato virus X and potato virus Y showed field resistance to both viruses and reduced losses (Kaniewski et al 1990). Field resistance varies between transgenic lines, depending on the source of coat protein gene (Table 1). Nucleocapsid protein-mediated resistance has been very effective in developing resistance to not only homologous but also heterologous strains of tomato spotted wilt virus (Pang et al 1992), which is emerging as the most serious pest in various parts of the world.

Integrated management of viral diseases in sustainable agriculture

No single approach can be effective in sustained management of the diseases caused by viruses. For example, use of acid-cured tomato seed reduces infection by ToMV but not very effectively because infection also occurs through soil debris, particularly in glasshouses. In the 1970s, classical cross protection in plants preinoculated with ToMV mutant M II-16 was highly successful in Europe, but plants grown at 25–30 °C were not protected, and under some environmental conditions the mutant reverted to virulent form (Brunt 1986). Viral coat protein-mediated resistance is an addition to the armoury of ToMV management. An integrated approach using clean seed, disinfected soil and resistant genotypes (developed through classical breeding or genetic transformation) can provide a more effective and lasting control of ToMV than any single approach.

Integrated management of vector-borne viruses is even more important, because the direct damage caused by vectors is also minimized. Of the various approaches that can be integrated for the management of different groups of viruses (Table 2), use of resistant varieties is most important because it requires fewest inputs from the growers. However, overdependence on this approach is not desirable because (a) it is difficult to develop resistant varieties with desirable traits and (b) unless inoculum pressure is restricted, the resistance is more likely to break down quickly.

TABLE 1   Incidence of tomato mosaic virus infection in transgenic tomato expressing the viral coat protein gene in a field trial in Florida in 1990

Tomato line	Coat protein gene source	% Plants infected Uninoculated 4 weeks	8 weeks	Inoculated[a] 4 weeks	8 weeks
UC 82 B	None	5.1	36.7	97.5	100.0
2068	Tobacco mosaic virus	0.0	0.0	5.1	50.8
3724	Tomato mosaic virus	2.5	10.3	17.5	60.0
4174	Tomato mosaic virus	0.0	0.0	0.0	20.0

[a]Plants were inoculated 17 days after transplanting and samples taken four and eight weeks later for ELISA.
From Sanders et al (1992).

TABLE 2 Management practices that can be integrated for the control of six selected groups of viruses

Management practice	Virus group WTG	Poty	Tobamo	Potex	Luteo	Tospo
Avoidance of sources of infection						
Use of virus-free true seed	x	+	+	?	x	+
Use of virus-free vegetative propagules	+	+	+	+	+	+
Avoidance of weeds, volunteers, overlapping generations, other crops, etc	+	+	+	+	+[a]	+
Use of virus-free soil and implements	x	x	+	+	x	x
Avoidance and control of vectors						
Avoidance	?	+	x	−	+	?
Chemical control	+	?	x	?	+	+
Biological control	?	+	x	−	+	?
Oil spray	?	+	x	x	?	?
Modification of cropping/ cultural practices						
Clean cultivation	+	+	+	+	+	+
Intercropping	?	+	+	+	+	?
Mulches	+	+	x	?	+	?
Resistance						
Resistant varieties	+	+	+	+	+	+
Induced resistance						
Cross protection	−	+	+	+	−	?
Systemic acquired resistance	−	?	?	?	−	?
Transformed plants	?	+	+	+	?	+

+, effective approach; −, ineffective approach; ?, efficacy doubtful; x, approach not applicable; WTG, whitefly-transmitted geminivirus. The groups of viruses are selected on the basis of their economic importance and transmission characteristics. Tobamoviruses have no specific vector and are transmitted by contact; potexviruses are transmitted by a fungus; potyviruses by aphids in a non-persistent manner; tospoviruses by thrips; luteoviruses by aphids and WTGs by whiteflies in a persistent manner.
[a]In sugar beet, weeds can be an advantage in the early stages of crop growth by providing alternative hosts for virus vectors and sources of predators and parasites (R. Plumb, personal communication, 1993).

Avoidance of source(s) of infection is ideal for the management of all diseases, but it is difficult to achieve, particularly for viruses with a wide host range. The most important step for any programme of integrated viral disease management is the use of virus-free planting material (Table 2). We must take advantage of the latest biotechnological developments for diagnosis of viral diseases in establishing and supporting programmes for the production and supply of virus-free planting materials. The use of healthy planting material must be supported by adequate management of the vector populations, which requires a good understanding of the ecology of viruses and their vectors, and reliable prediction models for timely action. It requires special efforts to generate location-specific data and prediction systems.

Management of viral diseases is possible only through integration of various approaches like those described in this paper. This has been best demonstrated in vegetatively propagated crops, such as potato and citrus, for which entire stocks and/or areas have been freed from viral and virus-like diseases through serious efforts. Given the benefits of technological developments, it is now possible to devise and implement similar programmes for all crops affected by viral diseases for sustainable agricultural production. In the foreseeable future, biotechnology is also expected to form an important part of integrated management of viral diseases for environmentally safe and improved crop production. Integrated management should not be left until after a pest reaches a level of economic importance. Viral diseases require a constant vigil and can best be managed when they first appear in an ecosystem.

Acknowledgement

I would like to thank Dr Roger Plumb for his critical suggestions during the preparation of this paper.

References

Aubert B, Quilici S 1984 Biological control of the African and Asian citrus psyllids (Homoptera: Psylloidea) through eulophid and encyrtid parasites (Hymenoptera: Chalcidoidea) in Reunion Island. In: Proceedings of the 9th Conference of the International Organisation of Citrus Virologists. IOCV, Riverside, CA, p 100–108

Bottenberg H, Irwin ME 1992 Using mixed cropping to limit seed mottling induced by soybean mosaic virus. Plant Dis 76:304–306

Brunt AA 1986 Tomato mosaic virus. In: Van Regenmortel MHV, Fraenkel-Conrat H (eds) The plant viruses, vol 2: The rod shaped plant viruses. Plenum, New York, p 181–204

Dahal G, Hibino H, Gabinagan RC, Tiangco ER, Flores ZM, Aquiero VM 1990 Changes in cultivar reaction to tungro due to changes in 'virulence' of the leafhopper vector. Phytopathology 80:659–665

Enyedi AJ, Yalpani N, Silverman P, Raskin I 1992 Signal molecules in systemic plant resistance to pathogens and pests. Cell 70:879–886

Fauquet C, Fargette D 1990 African cassava mosaic virus: etiology, epidemiology and control. Plant Dis 74:404–411

Fraser RSS 1987 Genetics of plant resistance to viruses. In: Plant resistance to viruses. Wiley, Chichester (Ciba Found Symp 133) p 6–22

Gibbs AJ, Skotnicki A 1986 Strategic defence initiatives; plants, viruses and genetic engineering. In: Varma A, Verma JP (eds) Vistas in plant pathology. Malhotra Publishing House, New Delhi, p 467–480

Hull R, Davies JW 1992 Approaches to nonconventional control of plant virus diseases. Crit Rev Plant Sci 11:17–33

Hunger RM, Sherwood JL, Evans CK, Montana JR 1992 Effects of planting date and inoculation date on severity of wheat streak mosaic in hard red winter wheat cultivars. Plant Dis 76:1056–1060

Kaniewski W, Lawson C, Sammons B et al 1990 Field resistance of transgenic Russet Burbank potato to effects of infection by potato virus X and potato virus Y. Bio/Technology 8:750–754

Kassanis B, Varma A 1967 The production of virus-free clones of some British potato varieties. Ann Appl Biol 59:447–450

Khetarpal RK, Maury Y, Cousin R, Burghofer A, Varma A 1990 Studies on resistance of pea to pea seed borne mosaic virus and new pathotypes. Ann Appl Biol 116:297–304

Lenne JM, Wood D 1991 Plant diseases and the use of wild germplasm. Annu Rev Phytopathol 29:35–63

Lindbo JA, Dougherty WG 1992 Pathogen-derived resistance to a potyvirus: immune and resistant phenotypes in transgenic tobacco expressing altered forms of a potyvirus coat protein nucleotide sequence. Mol Plant-Microbe Interact 5:144–153

Malathi VG, Varma A, Nambisan B 1989 Detection of Indian cassava mosaic virus by ELISA. Curr Sci 58:149–150

Pang SZ, Nagpala P, Wang M, Slighton JL, Gonsalves D 1992 Resistance to heterologous isolates of tomato spotted wilt virus in transgenic tobacco expressing its nucleocapsid protein gene. Phytopathology 82:1223–1229

Quemada HD, Gonsalves D, Slighton JL 1991 Expression of coat protein gene from cucumber mosaic virus strain C in tobacco: protection against infections by CMV strains transmitted mechanically or by aphids. Phytopathology 81:794–802

Rajamony L, More TA, Seshadri VS, Varma A 1990 Reaction of muskmelon collections to cucumber green mottle mosaic virus. J Phytopathol 129:237–244

Raychaudhuri M, Varma A 1983 Effects of oils on transmission of marrow mosaic virus by *Myzus pericae* Sulz. J Ent Res 7:107–111

Sanders PR, Sammons B, Kaniewski W et al 1992 Field resistance of transgenic tomatoes expressing the tobacco mosaic virus or tomato mosaic virus coat protein genes. Phytopathology 82:683–690

Sharma SR, Varma A 1982a Control of yellow mosaic of mungbean through insecticides and oils. J Entomol Res 6:130–136

Sharma SR, Varma A 1982b Control of virus diseases by oil sprays. Zbl Mikrobiol 137:329–347

Sharma SR, Varma A 1984 Effect of cultural practices on virus infection in cowpea. J Agron Crop Sci 153:23–31

Supriyanto A, Whittle 1991 Citrus rehabilitation in Indonesia. In: Proceedings of the 11th Conference of the International Organisation of Citrus Virologists. IOCV, Riverside, CA, p 409–413

Thresh JM 1981 Pest pathogens and vegetations. Pitman, London

Tien P, Zhang X, Qiu B, Qin B, Wu G 1987 Satellite RNA for the control of plant diseases caused by cucumber mosaic virus. Ann Appl Biol 111:143–152

Vani S, Varma A, More TA, Srivastava KP 1989 Use of mulches for the management of mosaic disease in muskmelon. Indian Phytopathol 42:227–235

Varma A 1976 Recent trends in control of plant virus diseases. Proc Natl Acad Sci India Sect B (Biol Sci) 46:193–206

Varma A 1984 Virus diseases of fruit and vegetable crops in Nigeria. FAO, Rome (FAO Tech Rep 10 AG:DP/NIR/72/007) p 101

Varma A, Sinha SK 1992 Sustainable development through long term biotechnological alternatives in agriculture. In: Daniel RR, Ravichandran V (eds) Proceedings of the International Seminar on Impacts of Biotechnology in Agriculture & Food in Developing Countries. Committee on Science and Technology in Developing Countries (COSTED), Madras, p 44–57

Varma A, Verma JP 1991 Progress and achievements in plant pathology. In: Srivastava US (ed) Glimpses of science in India. National Academy of Sciences India, Allahabad p 147–172

Varma A, Khetarpal RK, Vishwanath SM et al 1991a Detection of pea seed-borne mosaic virus in commercial seeds of pea, and germplasm of pea and lentil. Indian Phytopathol 44:107–111

Varma A, Khetarpal RK, More TA 1991b Genetical resistance in plants to viruses. In: Programme of the golden jubilee symposium on genetic research and education: current trends and the next fifty years. Indian Society of Genetics and Plant Breeding, IARI, New Delhi, p 127–128 (abstr)

Varma A, Dhar AK, Mandal B 1992a Mung bean yellow mosaic virus—the virus its vector and their control in India. In: Green J (ed) Mung bean yellow mosaic virus. Asian Vegetable Research and Development Center, Taipei, Taiwan, p 8–27

Varma A, Krishnareddy M, Malathi VG 1992b Influence of the amount of blackgram mottle virus in different tissues on transmission through the seeds of *Vigna mungo*. Plant Pathol 41:274–281

Vogel R, Bove JM, Nicoli M 1988 Le programme français de sélection sanitaire des agrumes. Fruits 43:709–720

Wetzel T, Candresse T, Macquaire G, Ravelanandro M, Denez J 1992 A highly sensitive immunocapture polymerase chain reaction method for plum pox potyvirus detection. J Virol Methods 39:27–37

DISCUSSION

Jeger: Over the last 15 years, it's quite striking how whitefly-transmitted geminiviruses have increased from being a relatively minor component of the literature on plant viruses to being possibly the most reported instance of serious epidemics throughout the world. This change in their occurrence and their incidence is observed in many different crops and agricultural systems, from Latin America and south-western USA to the Indian subcontinent. This is somewhat surprising, because geminiviruses aren't transmitted through seed; there is relatively limited systemic ability in some crops and there is only one vector, *Bemisia tabaci* (although adaptations or strains seem to be present in the vector and there may be other vector species). Why do you think this is happening and what do you think are the implications for sustainability, given that it seems to be occurring in both changing and relatively stable systems?

Varma: A major reason for the increased incidence of whitefly-transmitted viral diseases is the change in temperature. Whiteflies thrive at relatively high temperatures. Recently, there has been a slight rise in temperature, which has favoured an increase in the whitefly population. The second reason is the intensive cropping systems that are being introduced. Earlier, in say North India, nothing was grown during the summer period; the fields were left fallow. Now, with the introduction of irrigation, short duration crops are grown during the summer, providing an ideal situation for the early build-up of whitefly and for the growth of weed hosts that harbour the viruses. There is also a greater incidence of mung bean yellow mosaic virus in fields where crops are grown in the summer than in fields left fallow (Varma et al 1992).

Nagarajan: The common usage of the synthetic pyrethroids, which is occurring in various cropping systems around the world, might have contributed to the increase in the whitefly population, and thereby that of the whitefly-transmitted viruses. We have no experimental evidence of this; it should be investigated.

Varma: The use of synthetic pyrethroids has definitely resulted in an increase in the whitefly population, especially in South India because of the cotton belt. The effect may be even greater in the United States, where a new biotype of *B. tabaci* has been identified, which has a wider host range and is resistant to many insecticides (Gil 1992).

Wightman: There has also been a change in whitefly biotype on peanuts in the southern states of America.

Jayaraj: Before the introduction of synthetic pyrethroids, whiteflies, particularly *B. tabaci* and a few other related species, were of only minor importance, except in disease transmission. They were not important as pests. But after the large-scale introduction of synthetic pyrethroids in many countries, including the US, India, Thailand, Sudan and Pakistan, we find that the plants have become more prone to attack by whitefly. The resistance base of the plants has been weakened; also, the natural enemies of whitefly have been eliminated. We find increased fecundity of the whitefly. 40–50 whiteflies may be found per leaf, compared with the threshold of five. When the vector population is so high, we find a burst of increased disease transmission. Now we have come to a stage where whitefly-transmitted diseases can no longer be checked without interfering with the chemical crop protection *per se.*

Mehrotra: One may blame pyrethroids for the resurgence of whitefly, but there has been an increased incidence of viral diseases transmitted by whitefly all over the world. Most of the whiteflies in India have not been tested for resistance to pesticides. In Europe and America, they are resistant to organophosphates, to pyrethroids, to carbamates and to organochlorines. I do not understand fully the situation we are observing in India. There was a very strong resurgence of whitefly in 1930, before pesticides were used. Why? I think it has something to do with climatic change.

Varma: And changes in the cropping systems.

Ragunathan: I have also observed the problem of severe infestation with whitefly on certain crops after the introduction of synthetic pyrethroids. Whitefly damage was much more pronounced in cotton and in tobacco. Many of my colleagues have observed the disappearance of the parasites and predators of whitefly in cotton/tobacco growing areas where pyrethroids were used repeatedly. Other pests, such as *Alternaria*, have also emerged as a major problem in cotton in fields that received repeated applications of synthetic pyrethroids.

Whenever a pesticide is introduced into an area, we should take into account various aspects of the cropping system and the impact of the pesticide on the naturally occurring beneficial species. Otherwise, we may damage the biopotential present in that area and thereby affect sustainability in crop production.

I support what Dr Varma said regarding the importance of correct diagnosis of plant viral diseases. About 15 years ago, we observed virus-like symptoms in imported groundnut germplasm. Because of a lack of a clear diagnosis, some of the material was released for planting in a post-entry quarantine area for further observation. Now, the use of modern diagnostic kits enables easy identification of several strains of viruses.

Swaminathan: Quite often in the field farmers are confused between external symptoms due to micronutrient deficiencies and those due to viral diseases. Sometimes a presumed viral disease can be easily cured by application of a little zinc or manganese. The extension services generally are not fully informed about viral disease problems. They are also often not fully conversant with the impact of micronutrient deficiencies.

Temperature is extremely important. Progress in potato production in India was stimulated by the development of a seed plot technique that involves raising a seed potato crop during months when mean temperatures do not favour multiplication of the viral vectors. From an annual production of about three million tons of potatoes 20 years ago, annual production has now reached 15 million tons! India now produces more potatoes than the whole of Latin America, the home of the potato.

This was possible largely because of the development of a method of healthy seed production during the aphid-free season. With irrigation, and with the introduction of photoinsensitive genes into new varieties of crop plants, which makes them period-fixed rather than season-bound as in the past, there is a possibility, through a careful study of meteorological factors, of fixing the optimum time of planting, particularly in plants where there are serious problems of viral diseases.

Varma: I agree. The use of disease-free planting material and avoidance of vectors through knowledge of the environmental factors that influence the vector population should provide rich dividends.

A very important issue is the movement of genes for susceptibility. Neither cowpea golden mosaic nor cotton leaf curl is transmitted through seeds and these diseases did not occur in India until the introduction of exotic susceptible germplasm. It was the susceptibility to the causal viruses that was brought in; the viruses were already present in the wild. This topic merits further consideration.

References

Gil RJ 1992 A review of the sweet potato whitefly in southern California. Pan-Pac Entomol 68:14–152

Varma A, Dhar AK, Mandal B 1992 Mung bean yellow mosaic virus—the virus its vector and their control in India. In: Green J (ed) Mung bean yellow mosaic virus. Asian Vegetable Research and Development Center, Taipai, Taiwan, p 8–27

Agriculture in Gloria Land

Manindra Pal

Gloria Land, Sri Aurobindo Ashram, Pondicherry 605 002, India

Abstract. A farming system has been developed on the Gloria Land farm at the Sri Aurobindo Ashram that uses purely organic materials and achieves yields comparable with or better than those on conventional farms under similar agroclimatic conditions. The stimulus for the conversion to organic farming came from observations of the toxicity of chemical pesticides and their apparent ineffectiveness in reducing the impact of pests and diseases. On the Gloria Land farm, a carefully integrated mixture of activities includes crop growing, animal husbandry, fish rearing and sericulture. Sufficient organic waste is produced to fulfill all the needs of the farm's crops. Energy is partially supplied by biogas produced on the farm. This system is economically viable and ecologically sustainable.

1993 Crop protection and sustainable agriculture. Wiley, Chichester (Ciba Foundation symposium 177) p 158–167

I would like to leave the discussion of modern scientific agriculture for a while and try to impart some understanding of the experiences of farmers in developing a system called 'Self-reliant and low external input agriculture'.

Time alone can prove whether or not this system is sustainable. Meanwhile, I would like to describe the agricultural system I have developed over the last 28 years on a 40 ha farm called 'Gloria Land' at the Sri Aurobindo Ashram in Pondicherry.

A barren piece of eroded land criss-crossed with gullies and ravines with absolutely no water was purchased for growing fodder for dairy animals. An eminent agriculturalist who was a former member of the Agricultural Planning Commission volunteered to develop it. After adopting scientific measures of soil conservation, such as contour bunding, we started farming according to his instructions.

Between 1963 and 1970 I did all the field operations, including sowing, plant protection and harvest. I observed certain problems, mainly in the form of crop pests and diseases. Also, every time I sprayed the crop with pesticides, not only did I fall ill and vomit profusely, but the fish in the nearby canal where I used to bathe would suffer too. Some of the insect-eating birds also died because of the spray. This was a stunning revelation to me, although the experts did not agree, that wherever chemical fertilizers and pesticides were used, the pest problems were worse.

I then quietly experimented on a research plot, with promising results. On a plot with no chemical fertilizers or pesticides, yields were 25% lower than on one treated chemically, but the overall economic input:output ratio was equal. After frequent experimenting, the carrying capacity of the soil gradually increased as the micro and macro flora and fauna of the area were restored.

When the green revolution in India was gathering momentum, ashram farmers asked The Mother if chemical fertilizers could be used in the ashram fields. Her emphatic reply was, 'No, No, No. The whole world is making a mistake; we are not here to repeat the same'. This encouraged me to continue experimenting.

In the 1970s, we were flooded with literature about modern agriculture from all over the world. This promoted the use of chemical fertilizers, pesticides and herbicides as part of a package of crop production with indiscriminate use of hybrid seeds. The farmers were confused and because the results in the long run were detrimental to crop production, we decided to use our own initiative, keeping in mind the spiritual truth of the interdependency of all life forms, to probe the inadequately explored field of organic agriculture. We had to work out a programme of transition from the conventional method of farming to a sustainable one, based mainly on recyling of crop by-products and other waste materials from the farm mixed with cow urine and biogas slurry, plus crop rotation with green manure.

We also worked on improving the quality of the soil and soil organisms. We were guided by the wisdom and experiences of our Indian poets and seers. For example, Nobel laureate Tagore once wrote:

One who has overlooked the fundamental truth about the vitality of the Soil on which life sustains cannot hope to survive for long. If man returns to the soil the elements of life which sustain him, then his commerce with the soil will continue smoothly. In deceiving the soil he only deceives himself. It is only when the balance sheet of the soil shows an inflated debit column on a prolonged span of time and an inadequate credit column, that we realise it won't be long before man has to declare himself insolvent. We are sons of the earth and are nourished by her bounty and if we do not replenish her in the same proportions our consumption would then destroy everything. If we do not bring forth new thoughts and endeavours, and pursue them sincerely, the ecosystem and social systems of man would become lifeless.

Therefore, unless we take care of soil life, the interdependency of life on earth cannot be sustained. If the soil life is disturbed in any way, the entire chain of life will be affected. In other words, such disturbance would be suicidal. For this reason, I tried to develop an integrated approach by harnessing all the by-products and waste materials of the farm. At Gloria Land we are now able to fulfil almost all our food requirements without the use of chemical fertilizers and chemical pesticides. Our yields per ha are equal to those achieved with the use of such chemicals in India under similar agroclimatic conditions.

We have proved over the last 20 years that the input:output ratio can be balanced in an integrated farming system that includes dairy produce, fishery, poultry, sericulture, apiculture and duck-keeping. Chemical fertilizers can be totally replaced by organic manure produced on the farm, by recyling all waste products. Energy needs are supplied by biogas. What is being described today about China's achievements in this regard was practised by the ashram farmers 20 years ago.

Some of the yields from Gloria Land are:

(1) Paddy: we regularly harvest 4–6 t per ha per crop (fine variety).

(2) Perennial fodder Hybreed Napier: we harvest a minimum of 400 t per ha (compared with 200 t per ha with chemical fertilizers). The same yield can be maintained for over 10 years in the same field.

(3) Brinjal and tomatoes: 25–37 t per ha per crop.

(4) Banana: 100 t per ha per year.

(5) Coconuts: an average of 200 nuts per tree per year.

(6) Oil seeds (sunflower and groundnut): 2500–3700 kg per ha per crop.

(7) Fruits (e.g. mango, guava, sapota, jackfruit) are also grown and harvested.

Gloria Land is also active in biological pest management. We use cow urine, herbs, neem cake, neem leaves, asafoetida (a resinous gum of various Persian plants of genus Ferula, which has a strong smell of garlic and a bitter taste, and is used medicinally), fresh cowdung solutions, etc, in different combinations for different situations. These have totally replaced chemical pesticides that destroy the ecosystem and disrupt the balance of nature.

Between 1986 and 1988, we have conducted national and international training courses in ecological agriculture to disseminate our knowledge to hundreds of workers from non-governmental organizations.

I am often asked: 'Where do you get so much organic matter?' The answer is simple: you have to generate it. All that is needed is proper planning of the transition. Besides farms in the ashram, many other farms all over the world have developed similar integrated systems that could easily serve as models and even as examples for scientific evaluation.

However much the advocates of chemical pesticides and fertilizers assert that the use of these substances is indispensable to farming, they are bound to fail in the long run because these agents are destructive by nature—they cater only to those elements and components that are needed immediately by the plant. They overlook the immense role of natural soil life. The many letters that we at Gloria Land receive concerning this issue is a clear indication that farmers are earnestly looking for an alternative.

In summary, we have tried to combine all the agricultural disciplines and have paved the way for optimizing crop and animal productivity to the benefit of our country, bearing in mind The Mother's message: 'None of the present achievements of humanity can be an ideal for us to follow'.

DISCUSSION

Kenmore: Mr Pal, you have developed a number of techniques over a long period of experimenting. Are you working with younger people on Gloria farm who would then do more experiments and carry on in the future? If so, how do you organize that, which in another system might be called the research component?

Pal: Every part of work at Gloria Land is research, but it has not been published. I thought that nobody appreciated this method, because on television or on the radio all the propaganda is for chemical agriculture. Where is the propaganda medium open to me? That is one reason our methods have not spread, but they must have become more widely known, otherwise I would not be invited to national and international conferences such as this.

We have 13–14 people working on the farm, but remember this farm is not mine, it belongs to an institution. We are restricted to employing paid workers who are not motivated. Motivated people who will carry out the work are a scarce commodity. Of course, there are many agencies who could provide funds, but again there are restrictions on whether we can use those funds for research or not. Because of the terrorist activities in our country, if any foreign funds are used, there will be many enquiries per day from various government quarters, monitoring their use.

With all these constraints, it was I, on behalf of Gloria Land and the Sri Aurobindo Ashram, who brought to India for the first time the Agriculture, Man and Ecology Foundation, which was started by ETC in Holland. In 1982 I went there as a student; in 1984 I went there as a guest lecturer. After that, they were so impressed by my work that they came to Pondicherry and a training centre was organized at Gloria Land. I myself conducted ecological agricultural training courses for two years. 250 participants have been trained. From the little feedback that I have, I know that at least 150 of them are practising our methods. Dr K.K. Bhattacharyya was taken on as senior agronomist in 1987. After that we had to shift the training centre from Gloria Land to outside, because of the use of foreign funds.

Jayaraj: With our students, we collected some data from the Gloria Land farm on the build up of beneficial microflora in the soil. Because of the continuous organic farming they practise, there were many beneficial organisms, which check the root pathogens of many crops. The organic carbon content was very high. They have been using the local varieties of crops, which didn't respond very well to high levels of fertilizers on other farms, with organic farming. The outbreaks of pests and diseases were very few. I noticed only occasionally some vector-borne viruses in the grain legumes, like mung bean, which could not be checked because the vectors, such as whiteflies, were migrating onto this farm and spreading the diseases. Fungal diseases were few and there were not many insect pests. The quality of rice produced on the farm was good. But the method is not spreading, because other farms don't have that quantity of organic manure. At Gloria Land they have a big herd of animals, they produce methane and they use the entire sludge mixed with water as irrigation. Elsewhere, the shortage of manure is a limitation.

Secondly, I don't know whether Manindra Pal has statistics, but this is not a highly productive area. We have to produce more food. To do this, we sacrifice everything, which is not sustainable. At Gloria Land sustainability is constant, but they do not achieve high productivity. So perhaps we will have to develop an alternative system in which there is judicious use of pesticides and fertilizers in the form of integrated pest management and integrated nutrient management, without polluting the environment, but at the same time without sacrificing the realizable yield potential.

Hoon: I would like to comment on the success of Gloria Land, their farming practices and some of the problems we have encountered when trying to implement those strategies of organic farming in the villages. There are three factors that make Gloria Land unique.

As Manindra Pal pointed out, there is a strong spiritual component at Gloria. This has led to the success of the farm. Gloria was created at the insistence of 'The Mother', the spiritual head of Aurobindo Ashram. She wanted an assured supply of food grown organically and free from pesticides. The people who manage Gloria are followers of The Mother and are strongly motivated to fulfil her desires.

Manindra Pal describes the farming system at Gloria as cow-centred farming. They have approximately 250 dairy cattle at Gloria that provide sufficient manure for the farmland. Thirdly, they have a ready market at the Aurobindo Ashram for all the food produced at Gloria.

Gloria Land is in Pillayarkuppam village. There are 40 ha that belong to Gloria; another 140 ha are shared among 370 families. 70% of these people do not own land; the rest are marginal and small farmers. A large proportion of the landless people are employed at Gloria farm.

Despite the proximity to Gloria and the fact that several of the villagers work at Gloria, we found that the farming practices in the village remain quite uninfluenced by the farming practices at Gloria. Each farmer owns about two cattle and the dung is simply not sufficient to fertilize three crops. Much of the land is under sugarcane, because a nearby sugar mill provides a ready market. Also, the spiritual component that the inhabitants of Gloria farm have is missing. I detected a sense of total apathy amongst the villagers. There is a feeling of futility and hopelessness. The suicide rate is high, as is the rate of alcohol consumption. These are people who have lost hope; this is where we face the main problem.

The integral faith that makes Gloria a success is intangible and cannot be preached by extension workers. It is something people have to experience for themselves. Despite the proof at Gloria that high yields can be maintained through organic practices and proper management, it is not easy to transfer similar crop protection and management practices to the farmers' fields.

Varma: Control of viruses has been mentioned a couple of times in relation to organic farming systems. This reminds me of a method that was used in England about 30–40 years ago. In glasshouses, tomato mosaic virus, a non-vector-borne virus, was a problem that could be controlled by spraying plants with cow's milk. Although conflicting results were reported by different workers (Broadbent 1964), this approach merits further enquiry to obtain a scientific explanation. Induction of pathogenesis-related proteins by cow's milk may impart systemic acquired resistance.

Rigveda and *Atharvaveda*, the old writings (1500–500 B.C.), also describe methods for the control of plant diseases. An ancient book, *Vraksha Ayurveda* (*vraksha*, tree; *ayurveda*, science of medicine) lists methods similar to those mentioned by Mr Pal, like the use of cow dung, cow's milk, tumeric and neem (Singh 1983). So these methods are very old and perhaps we should look into their scientific basis. Dr Jayaraj said there are many beneficial microflora in the soil at Gloria Land farm; vesicular-arbuscular mycorrhizal fungi also increase in such systems.

Savary: Paul Teng is publishing a book on the traditional pest control practices in the Philippines. The book addresses not only the practices but some scientific explanation of how they work.

Saxena: The reason neem continues to be effective as an insecticide, in contrast to synthetic compounds, is that it affects insects in 12 different ways. 12 different modes of action have been identified.

Swaminathan: One can find scientific explanations, but the fact remains that Gloria Land is one of the few organic farms I have seen where complete ecological agriculture is practised with a high yield level. We always worry whether when ecologically desirable practices are adopted, productivity will fall. Mr Pal did experience a dip in productivity at the beginning, then he developed a method to restore the fertility of the soil, maintain adequate crop nutrition and ensure integrated pest management.

Bhattacharyya: I'm well acquainted with the system at Gloria Land, because I stayed there for four years. In this kind of farming system, transition planning is very important. I have seen at least 19 ecological farms spread over south India—in Karnataka, in Tamil Nadu and in Kerala—and the transition planning is crucial. If we try to reduce external inputs in a single step, the farmers will meet with disaster. That is the experience of many of the ecological farmers. It has to be done in phases, very carefully, because, particularly in India, no farmer can afford to lose yield and income. I know Manindra Pal changed his farm in a phasic manner.

Neuenschwander: My work is in biological control, which is only a small component of what is going on at Gloria Land, but I can support what Dr Bhattacharyya said. Biological control works very slowly. It takes sometimes years to reach a stable level. Dr Jayaraj suggested that we might combine biological control with regular use of some insecticides, but I wonder whether this is not chasing after the impossible. In many IPM systems relying on regular use of insecticides, applications of chemicals were reduced gradually, but biological control didn't perform very well. But did we give the natural enemies a good chance? I think that often we did not. There might be no low level of repeated insecticide use that doesn't throw us back to the zero point where biological control simply doesn't work.

Kenmore: One of the important elements in assessing the transition to biological control is to have adequate instrumentation. When we started using more sensitive ways to monitor changes in the density of rice field herbivores and changes in the age structure of the population as opposed to just taking an average density, we found that change happened incrementally, but steadily. Within one month there was measurable change—either when removing fields from a pesticide umbrella or by introducing pesticides into a previously untreated field. The changes were very quick and they tended to happen in terms of age structure rather than density. Naturally, overall population density would change after the age structure changes.

Adequate instrumentation means getting some of the things that were formally considered researchers' tools, including the *concept* of age structure, into the hands of people who are otherwise called extension workers, or people in NGOs, or farmers. Their new ownership helps them reduce pesticide use and move to a more sustainable farming system. But you have to give them the instruments so they can see the impact, their positive feedback, quickly.

Rabbinge: You have to distinguish between reducing the overuse of pesticides and what we have seen in many of the overintensified systems where biological self-reliance has virtually disappeared. You should try to promote biological self-reliance by giving the biological elements a chance to recover and by eliminating some of the crop growth-reducing factors. All IPM is directed towards this.

On the other hand, there is organic farming. If you are aiming at high yields per unit area, because of the number of people you would like to feed, then it's very important to see how open or closed the system is. Open or closed in the sense of inputs, in terms of energy or in other terms. How many people are able to live on the Gloria Land farm? How many people is it possible to support per hectare from such a farm?

Pal: 16 people from the ashram work in Gloria Land, and at present we are responsible for the management. In addition, about 80 workers come from the village regularly, although we are capable of managing the farm with only 20 workers. Using the mechanization we have so far, I could manage with 10 workers, but I prefer to absorb some of the unemployment problem in the local community. This farm is run commercially. For every commodity that is supplied to the ashram, money is paid according to the market price. All the development, including the infrastructure, has been paid for from our own earnings. 700 000–800 000 rupees per year in the form of labour wages goes to the landless people in the village. The food we are producing meets the requirement of 750 people in the ashram, plus about 700.

von Grebmer: Is there any controlled study comparing chemical farming, integrated pest management farming and organic farming, and measuring certain parameters over time, for example soil organisms, in the Third World.

Zadoks: Not in the Third World.

von Grebmer: I get the feeling we are discussing beliefs and convictions, which lack empirical data and quantification. It might be totally irrelevant later, but some quantitative, measurable indicators would be useful.

Bhattacharyya: In the Agriculture, Man and Ecology programme, we are comparing conventional farms with ecological farms in terms of economics and sustainability. We have three years' data for only seven pairs of farms: six in Tamil Nadu and one in Karnataka. So far, in terms of yield, ecological farms are equally productive: economically, ecological farms are better. It needs a few more years before we can draw firm conclusions. We are also measuring the build up of microflora and fauna in the soil.

Zadoks: There is a preliminary report on this group of farms from the Dutch Agricultural Economics Institute (van der Werf 1992).

Varma: At the Indian Agricultural Research Institute, we have started an experiment to compare the long-term effects of organic and advanced cropping methodologies for three cropping systems based on rice, pigeon pea and cotton.

Saxena: On the basis of my 16 years experience with rice in the Philippines, and now at a very remote place in Africa, at Mbita Point by Lake Victoria, a concept is emerging in my mind. This is that integrated pest management, which has been the most popular of the modern pest management systems for about 30 years, must itself undergo some evolution. Modern pest management has to be more biologically intensive. Instead of IPM, we should start calling it BPM—Biointensive Pest Management. This Biointensive Pest Management is very flexible and very dynamic (Fig. 1).

The IPM system developed at IRRI was dominated by the use of insect-resistant rice strains. Host plant resistance forms the hub of this system of pest management. The central focus could be taken over by biological control if one is dealing with cassava as the crop and green spider mite as the pest. For rice, the BPM wheel is also propelled by other biological components: biological control agents; botanical pest control, which incorporates some of my own experience with things like neem and other non-edible oil seed trees; cultural practices; biorational control, comprising the use of microbials and pheromones; supportive tactics, including pest monitoring, sampling, etc. We still have insecticidal control, but only as the last resort, or it may be dispensed with altogether.

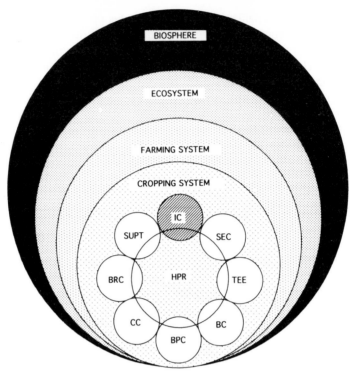

FIG. 1. (*Saxena*) The concept of biointensive pest management emphasizes sustainable ways of limiting the effects of pests on crop productivity and of improving crop health. It recommends maximal use of naturally occurring plant resistance, natural products, disease agents, predators, parasitoids and cultural practices. The major research imperatives for crop health are: (1) Cultural control (CC) based on better understanding and management of crop patterns, crop mixtures, plant residues and soil. (2) Maximizing intrinsic host plant resistance (HPR) to pests. (3) Enhancing the activity of natural biological control (BC) agents through conventional and new biotechnological methods. (4) Using safer botanical pest control (BPC) agents derived from locally available flora. (5) Biorational control (BRC) using hormones and chemicals that modify behaviour and using microbial insecticides. (6) Using chemical insecticides (IC) with improved methods for selective application. (7) Improved decision making, support for action and systems design (SUPT) based on regular monitoring and data collection concerning populations of pests and their natural enemies, the extent of crop loss and assessment of the economic damage. (8) Consideration of local socio-economic constraints (SEC). (9) Training in ecological education (TEE). The biointensive pest management system fits within broader ecological systems, as shown.

All this has to be done in cooperation with the communities in the area in which one is operating. At Mbita Point, in Kenya, there is no insectidal control, because the farmers have no cash, they cannot afford insecticides.

Then we have to add onto this BPM wheel the concept of training in ecological education. Mr Pal mentioned the interdependence of life. The tree cover is decreasing in most areas. In the Philippines and in Africa I see a lot of denuded and eroded lands. BPM also has to fit within a cropping system, and the cropping system has to fit in a farming system. Where I live, the cattle, goats, sheep and chicken are very important. They are an important source of protein to the resource-limited farmer. Eventually, the farming system itself is a component of an ecosystem. Lake Victoria is a very important ecosystem: it generates 400 million Kenyan shillings per year. In the Philippines, Lake Los Banos is an important source of fish, but its productivity has been falling because of heavy pollution by agrochemicals. The whole ecosystem fits in the larger system of the biosphere. Thus we see that the IPM concept itself is evolving and becoming more biologically intensive.

We hope that future IPM strategies will be ecologically oriented rather than pesticide driven. IPM practitioners, agricultural producers, the pesticide industry, food processors, governmental and non-governmental organizations and consumers will all have to share the responsibility of developing and implementing BPM.

References

Broadbent L 1964 Control of plant virus diseases. In: Corbett MK, Sisler HD (eds) Plant virology. University of Florida Press, Gainesville, FL, p 330–364

Singh RS 1983 Plant diseases. Oxford & IBH, New Delhi

van der Werf E 1992 Can ecological agriculture meet the Indian farmer's needs? In: Hiemstra W, Re'ntjes C, van der Werf E (eds) Let farmers judge. Experiences in assessing the sustainability of agriculture. Intermediate Technology Publishers, London

The local view on the role of plant protection in sustainable agriculture in India

S. Jayaraj and R. J. Rabindra

Tamil Nadu Agricultural University, Coimbatore, Tamil Nadu 641 003, India

Abstract. Indiscriminate use of chemical insecticides has affected humans and their environment and contributed significantly to reduced productivity of crops. With the increasing realization of the importance of sustainable agriculture, the concept of integrated pest management (IPM) for sustainable agriculture has emerged. In the recent past entomologists and the farmers have identified methods of pest management that are ecologically non-disruptive and stable. Concurrently indigenous crop varieties with resistance to pests and diseases have been developed and cultivated. According to the principle of 'organic farming', several non-chemical methods have become popular among the local farmers. Simple cultural practices like increasing the seed rate to compensate for pest damage, adjusting the time of sowing to avoid pest damage, mulching, intercropping, trap cropping and crop rotation have been found to provide adequate protection from pest damage with no additional cost and without harmful effects on the environment. The age-old method of catch and kill is still being practised by farmers, particularly for cotton. Mechanical methods like the bow trap for control of rats and provision of tin sheets around coconut tree trunks to prevent rats damaging the nuts are still being adopted. The use of botanical materials such as the neem products for pest management has been well received almost all over the world. Biological control using the natural enemies of insect pests has become very popular among the farmers in the 1980s. The farmers who clamoured for chemical pesticides in the 1960s and 1970s are now disillusioned with these poisonous eco-destabilizing substances; they want sensible, biologically rational methods of IPM. Pest surveillance and monitoring play an important role in IPM for sustainable agriculture.

1993 Crop protection and sustainable agriculture. Wiley, Chichester (Ciba Foundation Symposium 177) p 168–184

The green revolution in India has led to a quantum jump in agricultural production. With the titanic human population explosion, the pressure on agricultural land has increased tremendously, with the result that productivity and not merely the total production is of prime concern now. In the process of increasing productivity, more and more synthetic chemicals like fertilizers, pesticides, herbicides, soil dressings, etc, were used, which began to affect not only the health of the soil and agricultural production but the whole environment. Indiscriminate use of chemical pesticides, particularly, has affected humans and their environment and contributed significantly to reduced productivity of crops.

Pests, a constraint in sustainable agriculture

After the green revolution, there has been a marked change in the pests and diseases that affect many crops. Pests like locusts, red-hairy caterpillars and brown plant hoppers occur regularly in certain parts of India, often resulting in heavy crop losses. Many key pests, such as the cabbage diamondback moth (*Plutella xylostella*), mustard aphid (*Lipaphis erysimi*) and gram caterpillar (*Helicoverpa armigera*), have developed resistance to chemical pesticides. Large-scale cultivation of rice IR-64 in Indonesia, which was resistant to several pests, diseases and abiotic stresses, led to outbreaks of the white-backed plant hopper, *Sogatella furcifera*, to which the variety was highly susceptible, resulting in total loss of the crop. Year after year, the problems with crop pests have increased considerably and have become a major constraint in sustainable agriculture in many countries. According to Atwal (1986), the estimated value of annual crop losses caused by pests in India was Rs.51 812 million (Table 1).

Increased use of chemicals in agriculture

The consumption of technical-grade pesticides rose from 7341 t in 1960 to 79 482 t in 1989–90. The demand for 1994–95 is predicted to be 107 864 t (David

TABLE 1   Estimated value of annual crop losses caused by pests

Crop	Production (in million tonnes)	% Loss (Average)	Value of loss (in million Rupees)
Field			
Wheat	41.5	10	5191
Rice	53.6	20	13 078
Maize, sorghum, millets	30.0	25	9150
Cotton	8.0 m bales	50	3400
Brassicas	2.1	35–73	613
Groundnut	7.3	10	2380
Cottonseed	2.6	7	500
Pulses	11.35	10	5000
Losses during storage	54.2	7	12 500
Total			51 812

From Atwal (1986).

TABLE 2   Consumption of technical-grade pesticides in India in metric tonnes

Year	Insecticides	Fungicides	Herbicides	Rodenticides	Others	Total
1960	6729 (92)	598 (7)		104 (1)		7341
1965	10 428 (86)	1480 (12)		268 (2)		12 176
1970	21 741 (92)	1774 (7)	15 (0.06)	80 (0.3)		23 610
1974–75	46 515 (81)	9600 (17)	715 (1)	250 (0.4)	150 (0.3)	57 230
1977–78	46 755 (80)	9685 (16)	1425 (2)	270 (0.5)	845 (1.5)	58 980
1985–86						67 218
1989–90	60 450 (76)	11 387 (14)	5710 (7)	1790 (2)	145 (0.2)	79 482
1994–95[a]	81 390 (75)	15 564 (14)	8125 (8)	2615 (2)	170 (0.2)	107 864

Figures in parentheses indicate per cent share (David 1992).
[a]Estimated.

1992) (Table 2). Since 1960, there has been a nearly 12-fold increase in the consumption of insecticides alone, though the area under cultivation did not increase much after 1960.

Of the different groups of pesticides consumed in India, insecticides constitute about 80% (Atwal 1986) (Table 3). Of the total quantity of pesticides produced, nearly 68% goes into agriculture; the rest is used in the public health, industry and household sectors. Cotton occupies only about 5% of the cropped area but accounts for 52–55% of pesticides. The other crops, such as rice, cereals, millets, pulses, oil seeds, horticultural and plantation crops, sugarcane and others, which occupy about 95% of the cropped area receive 45–48% of the pesticides (Atwal 1986) (Table 4). Rice, cotton, fruits, vegetables and plantation crops together account for 89% of the consumption of pesticides, though they are grown on only 34% of cultivable area. Problems associated with pesticides are more frequent in these crops, resulting in less sustainable production.

To feed future generations without degrading the resource base that supports crop productivity, agriculture must become economically viable and ecologically sustainable. Through organic farming, sustainability could be achieved. It is not a question of eliminating chemical fertilizers and pesticides from the agricultural production scenario. The issue is how sufficient productivity can be achieved through integrated nutrient management (INM) and integrated pest management (IPM) using local resources, aiming towards sustainability.

TABLE 3 Relative amounts of different pesticides sold

Pesticide type	Market share	
	World (%)	India (%)
Herbicides	45	7
Insecticides	36	80
Fungicides	17	10
Others	2	3

From Atwal (1986).

Biodiversity

The biological wealth of India is considerable, with about 45 000 plant and 65 000 animal species. Silent Valley on the Western Ghats and the adjoining regions alone have contributed 20 major genes for disease and pest resistance in rice. About 152 plant species of economic importance originated in India. Agricultural production is more sustainable in areas where proper mixes of crops and varieties are adopted in a given crop season. Monocultures and overlapping crop seasons are more prone to severe outbreaks of pests and diseases. The diversity of natural enemies that attack pests at various stages of their life cycle is greater in mixed crop and intercropping systems, which also tends to prevent severe pest outbreaks.

Harmful effects of chemical pesticides

The use of synthetic organic pesticides has undoubtedly played a key role in increasing agricultural production in India. But indiscriminate use and abuse of these compounds have led to several toxic hazards through pesticide residues in the food chain (Kalra & Chawla 1986, Mehrotra 1986).

TABLE 4 Comparison of the amount of pesticide used and the area under cultivation for some major crops

Crop	Pesticide share (%)	Cropped area share (%)
Cotton	52–55	5
Rice	17–18	24
Chillies/vegetables/fruits	13–14	3
Plantation crops	7–8	2
Cereals/millets/oil seeds/pulses	6–7	58
Sugarcane	2–3	2
Others	1–2	6

From Atwal (1986).

TABLE 5   Pesticide resistance in pests of agricultural crops and stored grains

Insect species	Year resistance first reported	Pesticide(s)
Tobacco caterpillar *Spodoptera litura*	1965	Lindane, Endosulfan, DDT, Pyrethrum, organophosphorous compounds
Diamondback moth *Plutella xylostella*	1968	DDT, Lindane, cyclodienes, organophosphorous compounds, pyrethroids
Gram pod borer *Helicoverpa armigera*	1986	Cypermethrin
Mustard aphid *Lipaphis erysimi*	1986	Endosulfan
Tribolium castaneum	1971	Lindane, Malathion, Dichlorvos, phosphine
Sitophilus oryzae	1973	Lindane, Malathion, phosphine
Rhyzopertha dominica	1976	
Oryzaephilus surinamensis	1976	
Dermestes maculatus	1978	Lindane
Trogoderma granarium	1979	Phosphine
Tribolium confusum	1988	Malathion

From Kalra (1991).

Apart from these problems, it is becoming increasingly difficult to manage pests through the exclusive use of chemical pesticides. Budiansky (1986) has reported that the number of insects which exhibited resistance to pesticides was seven in 1935 and had swelled to 462 in 1986 all over the world. The cost of pesticides has multiplied 2000 times compared to in the 1950s. Repeated use of chemical pesticides tends to select pest populations resistant to them (Table 5), tempting the farmers often to double the dose and frequency of application and aggravate the already existing environmental problems. Several insecticides themselves are known to induce the resurgence of pests on crops (Jayaraj 1987) (Table 6). Because most of the chemical insecticides are also highly toxic to the natural enemies of insect pests, the use of such chemicals upsets the balance of life in nature, leading to large-scale outbreaks of pests.

The problems of resurgence of many minor pests are experienced all over the world (Jayaraj 1987). Jayaraj & Rangarajan (1987) have reviewed the situation with reference to the problems encountered after the introduction of synthetic pyrethroids.

TABLE 6   **Cases of resurgence of sucking pests induced by pesticides**

Pest		*Pesticide causing resurgence*			
		OC	OP	C	SP
Cotton					
Whitefly	*Bemisia tabaci* Gen.	1	3	3	3
Spider mite	*Tetranychus urticae* (Koch)	1	—	1	—
Red spider mite	*T. cinnabarinus* Boisd	1	1	1	6
Aphid	*Aphis gossypii* G.	1	1	1	8
Leaf hopper	*Amrasca biguttula biguttula* Ishida	—	—	—	1
Mealy bug	*Ferrisia virgata* (Cockerell)	—	—	—	4
Rice					
Brown plant hopper	*Nilaparvata lugens*	—	19	2	6
White-backed plant hopper	*Sogatella furcifera*	—	—	—	3
Green leaf hopper	*Nephotettix virescens*	—	1	—	1
Blue leaf hopper	*Emposacanara (Zygina) maculifrons*	—	6	—	—

C, carbamate; OC, organochlorines; OP, organophosphorous; SP, synthetic pyrethroids.
From Jayaraj (1987).

Pesticide application techniques

Apart from the development of resurgence and resistance, the failures to control pests in the field are also due to defective pesticide application technology. Recent surveys by scientists at Tamil Nadu Agricultural University (TNAU) have indicated that most appliances are far from satisfactory at delivering the target dose to the target surface. For example, effective control of the brown plant hopper could be achieved by applying the pesticide so that it reaches the culm. But none of the available appliances could deliver the pesticide to the culm in a rice crop. Most of the applications by farmers cover the upper surface of leaves, whereas pests like whiteflies, aphids and leaf hoppers are harboured on the lower surface of leaves. The wrong choice of chemicals and incorrect doses also aggravate the problems.

Alternatives to chemical pesticides

With the increasing realization of the importance of sustainable agriculture, the concept of IPM is well conceived for sustainable agriculture. Recently, entomologists and farmers have identified IPM methods that are ecologically non-disruptive and stable.

Resistant varieties of pests

Indian farmers have become so frustrated with the pesticide syndrome that they are clamouring for pest-resistant or pest-tolerant crop varieties. They are very aware that chemical insecticides constitute a major part of the cost of cultivation and that productivity and sustainability can be substantially increased by growing pest-resistant varieties which will reduce their reliance on insecticides.

Through intensive breeding programmes, the TNAU has developed several crops resistant to pests (Jayaraj & Uthamasamy 1990), some of which are presented in Table 7. Rice varieties resistant to particular biotypes of brown plant hopper and cultivars with different resistance genes have been developed for effective management of the pest in many rice-growing countries. Also, rotational use of varieties with different known resistance genes could make the varieties stable in the field by slowing the development of resistant pest biotypes (Chelliah & Bharathy 1985). The development of TKM 6, which possesses multiple resistance to several rice pests, is a notable achievement. This variety has been used as a donor in the development of many national and international varieties currently in cultivation.

Cultural methods

The farmers are well aware that by adopting or modifying certain cultural practices, the pest population and damage caused can be reduced. Increasing seed rate to compensate for shoot fly damage to sorghum, adjusting the time of sowing of cotton to reduce pest attack, and trashing and mulching in sugarcane to prevent pest damage are some popular local practices. Others include judicious application of nitrogenous fertilizers for management of plant hoppers in rice (Uthamasamy et al 1983), and the use of coal tar-coated urea or neem cake-blended urea to reduce infestation with the sucking pests, stem borer and leaf folder, in rice (Anonymous 1986). The application of *Azospirillum* biofertilizer helps control the sorghum shoot fly *Atherigona varia soccata* (Mohan et al 1987). Intercropping is another popular cultural practice: mung bean in cotton for reducing the leaf hopper population (Rabindra 1985), lab-lab in sorghum to reduce damage by the stem borer, *Chilo partellus* (Mahadevan & Chelliah 1986), pearl millet in groundnut to reduce damage by the leaf miner, *Aproaerema modicella*. Castor is grown as a trap crop in cotton, tobacco and chillies to reduce damage by *Spodoptera litura* (Jayaraj & Santharam 1985).

TABLE 7   Insect-resistant crop varieties in Tamil Nadu

Insect	Crop and variety
	Rice
Brown plant hopper	Co 42, ADT 36, PY 3, ADT 37
Green leaf hopper	TKM 6, ASD 5, ADT 37, ADT 38, PY 3[a]
Yellow stem borer	TKM 6
Gall midge	TKM 6, MDU 3, GEB 24
Leaf folder	TKM 6, GEB 24, TKM 2
	Cotton
Stem weevil	MCU 3
Leaf hopper	MCU 5, MCU 9
Whitefly	LPS 141, LK 861, Supriya
Bollworms	Abhadita
	Sorghum
Shoot fly	Co 1
Rice weevil	Co 4, Co 18, Co 19
	Sugarcane
Scale	Co 8014, Co 617, Co 6501
Internode borer	Co 283
Shoot borer	Co 281, Co 6601
	Brinjal
Aphid	Annamalai
	Jasmine
Eriophyid mite	Pari mullai

[a]Moderate resistance only.
From Jayaraj & Uthamasamy (1990).

Mechanical methods

One of the oldest methods of pest management is mechanical control. Hand collection of insects like *H. armigera* and *S. litura* on cotton, grain legumes, tobacco; hooking out *Oryctes rhinoceros* beetles from coconut crown; destruction of grubs in manure pits; setting bow traps in rice fields and providing tin sheets around coconut trunks for rat management are still practised by farmers. The TNAU has developed a suction device using the motorized backpack sprayer for mechanical removal of sucking insects like whiteflies (*Bemisia tabaci*), leaf and plant hoppers and aphids.

Physical methods

Activated clay at 1% was found to control effectively the storage pests of grain legumes and sorghum (Swamiappan et al 1976). Similarly, vegetable oils at 1% effectively reduced the damage by the pulse beetle, *Callasobruchus chinensis*.

Behavioural methods

Farmers have been taught to use pheromone traps for monitoring pests like the pink bollworm, *Pectinophora gossypiella*, *H. armigera* and *S. litura* (Jayaraj et al 1986). The use of these traps, as well as light traps in rice, in pest management is common among Indian farmers. Similarly, fish meal traps for *A. varia soccata* (Rangarajan et al 1985), and yellow pan traps or yellow sticky traps (previously used to monitor whiteflies) (Jayaraj et al 1985) are useful in pest management. Spreading blue cloth to attract the larvae of *S. litura* then destroying them is a novel method of pest management developed from farmers' practice (Dhandapani et al 1986). Such behavioural methods are cheap and do not disrupt the ecosystem.

Use of plant products

Indian farmers used pesticides of plant origin long before synthetic chemical pesticides were discovered. The plant kingdom is a vast storehouse of chemicals that can check the insect population. Grainage et al (1985) compiled a database of plant species that possess pest-controlling properties. This included 1005 species with insecticidal properties and 384 species with antifeedant properties; 297 were repellent, 27 were attractant and 31 species had growth-inhibiting properties. The use of neem and *Vitex negundo* products for pest management, both in the field and in storage, is quite popular.

Biological control

It is well known that natural enemies of insect pests play a key role in biotic balance, reducing pest populations below levels with an economic impact. Both natural and applied methods of biological control are important in successful management of pest populations. After nearly two decades of intensive teaching and field training, farmers have understood the value of biological control. Having realized that most of the synthetic chemicals decimate the beneficial parasites and predators, the farmers now tend to use selective pesticides along with biocontrol agents and botanical pesticides.

In the area of applied biological control, significant progress has been made. Since the successful biological control of the apple woolly aphid and cottony cushion scale, several entomophages and entomopathogens have been developed

TABLE 8 Bicontrol agents found effective in pest management

Natural enemy	Host insect pest
Parasites	
Trichogramma spp.	Sugarcane internode borer
Chelonus blackburni	Bollworm
Tetrastichus israeli	Coconut caterpillar
Eriborus trochanteratus	Coconut caterpillar
Goniozus nephantidis	Coconut caterpillar
Sturmiopsis inferens	Sugarcane shoot borer
Predators	
Cryptolaemus montrouzieri	Grapevine mealy bug
Scymnus coccivora	Grapevine mealy bug
Chrysoperla carnea	Aphids, whiteflies and young larvae of Lepidoptera
Epiricania melanoleuca	Sugarcane pyrilla
Lycosa pseudoannulata	Brown plant hopper
Platymeris laevicollis	Coconut rhinoceros beetle
Pathogens	
Nuclear polyhedrosis virus	*Helicoverpa armigera*
	Spodoptera litura
	Amsacta albistriga
	Plutella xylostella
	Spilosoma obliqua
	Hyblea puera
Granulosis virus	*Chilo infuscatellus*
Non-occlusion virus	*Oryctes rhinoceros*
Metarhizium anisopliae	*O. rhinoceros*
	Nilaparvata lugens
Beauveria bassiana	*N. lugens*
Verticillium lecanii	*Coccus viridis*
Bacillus thuringiensis	*P. xylostella*

From Jayaraj & Rabindra (1992).

for use in pest management (Table 8). The role of biopesticides, particularly microbial pathogens in biorational pest management, has been well documented (Jayaraj 1985, 1986, Jayaraj et al 1992, Rabindra et al 1992). The efficacy of microbial pathogens can be enhanced further by genetic improvement (Jayaraj & Rabindra 1992). The value of biocontrol is now well recognized, especially in the context of environmental protection as well as stable pest management.

Biological control programmes in Tamil Nadu state

The state of Tamil Nadu is a pioneer in implementing biological control programmes for pest management. The government of Tamil Nadu has set up a network of 42 parasite breeding centres as well as a few insect pathogen multiplication centres to cater to the needs of the farmers, particularly coconut, sugarcane, groundnut and cotton growers. Most farmers are well aware of the principle of biological control and can identify friendly insects on the crop plants. The sugarcane and coconut growers regularly release parasites for pest management.

The Government of India has sanctioned Rs.5000 million recently for six programmes including biological control in IPM, of which the allocation for the state of Tamil Nadu is about Rs.520 million. A major portion of this is for upgrading the existing parasite breeding centres to cover all crops. Under this programme, there will be an apex laboratory at the TNAU main campus and four regional laboratories at the Agricultural Colleges and Regional Research Stations. There will be 5–6 centres in each district in collaboration with the Directorate of Agriculture. This joint endeavour will be a model for other states and countries.

The TNAU has recently set up large-scale facilities for mass production of biocontrol agents under a programme supported by the Department of Biotechnology and the Biotech Consortium of India Ltd. The farmers now know that apart from parasites and predators, insect pathogenic viruses can be used as microbial pesticides. Being convinced of the efficacy of nuclear polyhedrosis viruses for the control of *H. armigera* and *S. litura* and the granulosis virus for the control of the sugarcane shoot borer, *Chilo infuscatellus*, farmers are now demanding these viral insecticides. Biological control has become so popular among the Tamil Nadu farmers that some agricultural graduates and farmers have set up commercial biological control laboratories.

Pest surveillance in sustainable agriculture

Pest surveillance and monitoring are essential to avoid unnecessary application of chemical insecticides. The TNAU and the State Department of Agriculture launched jointly in 1984 a pest and disease monitoring programme: nearly 7250 points in the state have adopted fixed-plot and roving surveys (Jayaraj 1988). The farmers are advised to take up pest management measures only when the pest population level exceeds the economic threshold. These are different for pests occurring along with their natural enemies from those when the pest occurs alone. The need-based application of pesticides according to the economic threshold has reduced the cost of cultivation. Estimates made by the State Directorate of Agriculture indicate an annual saving of nearly Rs.20 million worth of pesticides since the introduction of this programme. The plant clinic centres established in the state have contributed significantly to this programme.

Integrated pest management for sustainable agriculture

IPM has a key role to play in sustainable agriculture. Many ecologically based IPM practices have been developed (Jayaraj et al 1988). Through collaborative efforts with the development departments of Agriculture, Horticulture and Plantation Crops, and Seed Certification, the TNAU has implemented IPM programmes in rice, sorghum, cotton, groundnut and coconut in about 100 Technology Development and Demonstration Centres spread over the entire state. This programme has without doubt increased the sustainability in agriculture in the areas of its operation. The concept of IPM for sustainable agriculture must be given greater emphasis throughout India and other countries. Research efforts should be concentrated on developing ecologically stable and economically viable pest management strategies with a view to stabilizing agricultural production in a sustainable manner.

Acknowledgements

The authors are grateful to Dr M. Swamiappan, Professor of Agricultural Entomology and Mr M. Muthuswami, Senior Research Fellow of the Department of Agricultural Entomology, Tamil Nadu Agricultural University for their assistance in the preparation of the manuscript.

References

Anonymous 1986 Report of the scientific workers conference, Tamil Nadu Agricultural University, 29–30 April, 1986. TNAU, Coimbatore

Atwal AS 1986 Future of pesticides in plant protection. In: Venkataraman GS (ed) Plant protection in the year 2000 AD. Indian National Science Academy, New Delhi p 77–90

Budiansky S 1986 Set a bug to catch a bug. US News & World Report, 13 October

Chelliah S, Bharathy M 1985 Rice brown planthopper, *Nilaparvata lugens* (Stal.). Status and strategies in management. In: Jayaraj S (ed) Integrated pest and disease management. TNAU, Coimbatore, p 34–53

David BV 1992 Pesticide industry in India. In: David BV (ed) Pest management and pesticides: Indian scenario. Namrutha Publications, Madras, p 225–250

Dhandapani N, Abdul Kareem A, Jayaraj S 1986 Studies on use of colour cloth to trap cotton leafworm, *Spodoptera litura* (F.) (Noctuidae:Lepidoptera). Cotton Dev 15:11–12

Grainage M, Ahmed S, Mitchell WC, Hylin JH 1985 Plant species reportedly possessing pest control properties: an EWC/UM database. Resource Systems Institute, East West Centre, Honolulu

Jayaraj S (ed) 1985 Microbial control and pest management. TNAU, Coimbatore, p 278

Jayaraj S 1986 Role of insect pathogens in plant protection. Proc Indian Natl Sci Acad Part B Biol Sci 52:91–107

Jayaraj S (ed) 1987 Resurgence of sucking pests—proceedings of a national symposium, Tamil Nadu Agricultural University, Coimbatore. TNAU, Coimbatore, p 272

Jayaraj S 1988 Plant protection research in Tamil Nadu Agricultural University 1960–1988. An annotated bibliography, vol 1. TNAU, Coimbatore

Jayaraj S, Rabindra RJ 1992 Recent trends in increasing the efficacy of biocontrol agents. In: Ananthakrishnan TN (ed) Emerging trends in biological control of phytophagous insects. Oxford IBH and Publishing Co., New Delhi, p 1–9

Jayaraj S, Rangarajan AV 1987 Trends in insect control—post-pyrethroids. Pesticides 21:11–25

Jayaraj S, Santharam G 1985 Ecology-based integrated control of *Spodoptera litura* (F.) on cotton. In: Jayaraj S (ed) Microbial control and pest management. TNAU, Coimbatore, p 256–264

Jayaraj S, Uthamasamy S 1990 Aspects of insect resistance in crop plants. Proc Indian Acad Sci Anim Sci 99:211–224

Jayaraj S, Rangarajan AV, Jeyarajan R 1986 Pest and disease surveillance in cotton. In: Jayaraj S (ed) Pest and disease management—oilseeds, pulses, millets and cotton. TNAU, Coimbatore, p 169–180

Jayaraj S, Rangarajan AV, Santharam G, Vijayaraghavan S 1985 Whitefly—a threat to cotton cultivation. In: Jayaraj S (ed) Pest and disease management—oilseeds, pulses, millets and cotton. TNAU, Coimbatore, p 185–188

Jayaraj S, Rangarajan AV, Rabindra RJ, Rajasekaran B, Balasubamanian M 1988 Studies on non-chemical methods for pest management in Tamil Nadu. In: Mohan Dass N, Koshy G (eds) Integrated pest control—progress and perspectives. Association for Advancement of Entomology, Trivandrum, p 70–86

Jayaraj S, Sathiah N, Sundarababu PC 1992 Biopesticides research in India: the present and future status. In: David BV (ed) Pest management and pesticides: Indian scenario. Namrutha Publications, Madras, p 144–157

Kalra RL 1991 Status of pesticide resistance in insects in India. Paper presented at ICAR/IRPM/USDA joint project development group meeting of management resistance with focus on *Heliothis* resistance management in India, 17 October 1991, Hyderabad

Kalra RL, Chawla RP 1986 Pesticidal contamination of foods in the year 2000 AD. In: Venkataraman GS (ed) Plant protection in the year 2000 AD. Indian National Science Academy, New Delhi, p 188–204

Mahadevan NR, Chelliah S 1986 Influence of intercropping legumes in sorghum on the infestation of the stem borer, *Chilo infuscatellus* Swinhoe in Tamil Nadu, India. Trop Pest Manage 32:162–163

Mehrotra KN 1986 Pest control strategies for 2000 A.D. In: Venkataraman GS (ed) Plant protection in the year 2000 AD. Indian National Science Academy, New Delhi, p 10–16

Mohan S, Jayaraj S, Purushothaman D, Rangarajan AV 1987 Can the use of *Azospirillum* biofertilizer control sorghum shootfly? Curr Sci 14:723–725

Rabindra RJ 1985 Transfer of plant protection technology in dry crops. In: Jayaraj S (ed) Integrated pest and disease management. TNAU, Coimbatore, p 377–383

Rabindra RJ, Sathiah N, Jayaraj S 1992 Efficacy of nuclear polyhedrosis virus against *Helicoverpa armigera* (Hbn.) on *Helicoverpa*-resistant and susceptible varieties of chickpea. Crop Prot 11:320–322

Rangarajan AV, Chelliah S, Jayaraj S 1985 Pest management in field crops and stored products. TNAU, Coimbatore, p 72

Swamiappan M, Jayaraj S, Chandy KC, Sundaramurthy VT 1976 Effect of activated kaolinitic clay on some storage insects. Z Angew Entomol 80:385–389

Uthamasamy S, Velu V, Gopalan M, Ramanathan KM 1983 Incidence of brown planthopper, *Nilaparvata lugens* Stal. on IR50 with graded levels of fertilization at Aduthurai. Int Rice Res News 8:13–14

DISCUSSION

Royle: Have you utilized the Indian Meteorological services with your surveillance systems?

Jayaraj: Yes, we have. We have established a supercomputer linked to all the agroclimatic zones through the network of the Indian Council of Agricultural

Research Stations and State Agricultural Universities. We also have state-level monitoring of weather factors. Weather changes are major components of the pest surveillance programme. We have quantified the influence of changing weather on impending outbreaks of pests and diseases. Dr Nagarajan is an authority on this. His book on the *Epidemiology of crop diseases* is read all over the world and he has helped in developing this concept in India.

Waibel: You have 7250 surveillance points. Could you tell us something about the approximate costs of this system, including the computer and other equipment?

Jayaraj: We didn't develop it as a separate project in terms of funding; it is part of our job. We have several research stations and plant health care centres in the university. We involved subject matter specialists in plant protection from the Department of Agriculture who operate at the state level, regional level, district level and at the village level. Over the last eight years we have trained them, using the available resources of manpower and funds. We have provided the computers, which are used not only for this but also for other purposes. So I cannot cost this programme because it's part of the job.

Varma: You mentioned the yellow traps: how many traps were set up and what was the reduction in the population of whiteflies?

Jayaraj: There were 25 traps per hectare for grain legumes, cotton and so on. The reduction in population of thrips and whiteflies was about 70%. The colour and height of the trap were standardized. Thrips were attracted more than the whiteflies, which could be a problem. If so many thrips were attracted that they covered the surface, the subsequent attraction of whiteflies was poor. The castor bean oil on the traps has to be changed frequently, but this oil is available in every village.

Varma: What is the exact design of the traps?

Jayaraj: They consist of an inverted tin nailed onto a small post. The tin is painted yellow and a polythene sheet tied around it. Over the polythene sheet we apply the castor bean oil. It was previously used as a monitoring device. One of our students wanted to test it as a control tactic by increasing the number of traps per unit area, and it worked well.

Neuenschwander: I am astonished that you need so few yellow traps per hectare. In the Mediterranean area, yellow traps are used against different fruit flies, on olives for instance, mainly as a monitoring device. Suppressing the fly populations requires many traps per tree. In studying the effect of the yellow traps in detail, we found that parasitoids of the fruit flies and of two olive scale insects were caught by the yellow traps 16 times as often as were fruit flies. Worse, we found about 50 times as many parasitoids of unknown origins killed by the yellow traps. So yellow traps are not necessarily specific.

Jayaraj: Tree crops may require a higher density of traps. We worked mostly on grain legumes, which are short crops, and cotton to some extent. We didn't notice any large-scale attraction of natural enemies; perhaps we should look for the predatory mites, we didn't separate them out. I will ask my colleagues to look into this.

Mehrotra: Yellow traps have been used successfully to monitor *Lipaphis erysimi*, mustard aphid, which is a major pest of mustard. As you pointed out, sometimes a large number of parasites and predators are also attracted and get stuck on the traps, but this is rare if the traps are used properly.

Rabbinge: I was very impressed by your talk; you seem to be getting products into use very successfully. The combination of scientific knowledge with simple methods such as the drums is very appealing. How are you implementing this? Can you tell us more about the type of training and extension activities that are being organized?

Jayaraj: We have in India monthly zonal workshops in which the extension workers of the district come to the State Agricultural University station or college for two days a month. They discuss the success and failures of the previous month, and then the impending problems of the next month. We are able to tell them what is appropriate for the season and the location. In addition, we have many structured training programmes, of one or several months, particularly in a field like plant protection which is highly specialized.

We also have conferences for plant protection workers in each state and at the all-India level. At these we work out all the policy issues, and guide the extension workers, NGOs and farmers through better policies. We are fairly successful. We have a Centre for Plant Protection Studies at Tamil Nadu Agricultural University, established in 1984, which is leading the whole plant protection programme of the Department of Agriculture. In states where the link between research and development is very good, you have success stories. My state is one of those success stories.

Jeger: Could you say something about how you maintain quality control in the mass production of the biocontrol agents, such as viruses? In particular, how do you control the genetic identity and integrity of the virus?

Jayaraj: We isolated different strains of nuclear polyhedrosis virus (NPV) from different parts of the country and selected one. We are dealing with only one strain in the biopesticide commercial pilot plant and avoiding all contaminations with other strains.

I should say something about the safety. We have generated a lot of data on the safety of NPV with respect to non-target organisms. We are sure that the organism with which we are dealing is very safe to non-targets.

Quality control is relatively simple because the viral polyhedra measure about 3–5 μm in diameter. By centrifugation we can remove the other organisms. We are using live insects for *in vivo* production: whenever the symptoms exhibited differ from the norm, we reject the whole batch. By and large, we are not having this problem. We have been doing this for nearly 20 years, and we have developed methods to avoid contamination and quality control problems.

Finally, we assess the number of polyhedra per ml of the fluid. We sell the commercial product. The Biotech Consortium of India gave us two years to prove the commercial worth of this project; we showed it in six months! Farmers from neighbouring districts come to Coimbatore to buy the NPV, which is comparable in price to insecticide. They find it very useful, and the Government of India has sent a team of consultants to promote this in private industries.

Mehrotra: The pesticide resistance problem, as far as agriculture is concerned, started in 1963, but it is only now that we are reaching a critical stage. *Plutella* is no longer controlled by any pesticide available in India. In and around Bangalore, *Bacillus thuringiensis* is being used.

In *Spodoptera*, multiple resistance to pesticides has already been acquired. Next, if we do not take care, will be *Helicoverpa*. This started with a bang in the 1987–1988 cotton season in Andhra Pradesh (Dhingra et al 1988). Technical advice was given by both specialists and industry representatives and pyrethroids were allowed to be used only for certain periods (60 to 120 days old crop). This reduced the level of resistance drastically. A similar methodology was recommended in Punjab when there was a *Helicoverpa* epidemic in the 1990–1991 cotton season (Mehrotra & Phokela 1992). I'm afraid that aphids and jassids will become resistant to pesticides next. It is therefore essential that IPM strategies are developed for various pests in different cropping systems.

With storage it is the same story. Pests have become resistant to insecticides and the fumigant, phosphine. Biological control will have to be used to manage these pests. Long ago, on purely ecological principles, a cheap village level storage structure was designed. This was known as the Pusa bin and worked well. It needs to be made more popular with farmers.

We are already in bad shape as far as public health is concerned. We are losing 25–30 million days of labour per year, mainly in rural areas. In India, two species of malaria vector, *Anopheles stephensi*, which is an urban vector, and *An. culicifacies*, which is a rural vector, have become resistant to all the pesticides used in the Public Health sector. IPM projects will have to be started in the public health sector. The Government of India has specified IPM as official policy, but in the public health sector we need it urgently. Because the use of pesticides in public health matters is done entirely by government agencies, it can be easily controlled.

The whole question of pesticide resistance has been reviewed several times with respect to India and the need for IPM is clear (Mehrotra 1989, 1990, 1991, 1992).

Nagarajan: The Government of India has now officially recognized that botanical agents as well as biological pesticides like *B. thuringiensis* can be registered and marketed. This is a very big strategy change for plant protection that follows much in-house debate. Quality control, marketing and many other aspects will be regulated.

The Indian Council of Agricultural Research and the Agricultural Universities believe that several other botanical agents, in addition to neem, have potential as insecticides and as fungicides. There are several aromatic plants, like *Occimum* spp, which have fungicidal properties. These botanicals may not have to compete with fungicides or insecticides, because they have their own defined market, for example as a mosquito repellent, leather preserver or against cockroaches. They are thermolabile and photosensitive, so they are not being targeted to compete with chemical pesticides. In the next five years, better formulations of botanical pesticides are expected to come on to the market.

The second aspect is that we have taken our results on biological control in a very strong way. Both the council and the Department of Biotechnology have produced very ambitious projects which Dr Jayaraj explained with regards to NPV. Once this technology has been standardized, it should be passed on to the entrepreneur, who will market it locally. This concerns not only NPV, but also *Trichoderma*. Demonstrations showed that on a 15 ha plot this is able to give complete freedom from soil-borne diseases in chickpea. Conditions for mass multiplication and usage of other parasitoids are also being standardized.

References

Dhingra S, Phokela A, Mehrotra K 1988 Cypermethrin resistance in populations of *Heliothis armigera*. Natl Acad Sci Lett (India) 11:123–125

Mehrotra K 1989 Pesticide resistance in insect pests: Indian scenario. Pestic Res J 1:95–103

Mehrotra K 1990 Pyrethroid resistant insect pest management: Indian experience. Pestic Res J 2:44–53

Mehrotra K 1991 Current status of pesticide resistance in insect pests in India. J Insect Sci 4:1–14

Mehrotra K 1992 Pesticide resistance in insect pests—Indian scenario. In: David BV (ed) Pest management and pesticides: Indian scenario. Namrutha Publications, Madras, p 17–27

Mehrotra K, Phokela A 1992 Pyrethroid resistance in *Helicoverpa armigera* Hubner V. Response of populations in Punjab on cotton. Pestic Res J 4:59–61

General discussion II

Reducing pesticide use in Asia and Africa

Mehrotra: I am apprehensive about increasing use of pesticides, both insecticides and herbicides, on the paddy crop in India. The latest figures I have are that 39% of the pesticide used in India goes to cotton, and 35% to paddy (David 1992). This is a disturbing trend.

In this country, there used to be an Operation Research Programme on IPM in paddy under the Indian Council for Agricultural Research. In 1978–1979 I visited Andhra Pradesh to evaluate this programme. In the fields where no pesticide was used, there were many mirid bugs, spiders and other parasites: this was not true for the fields that had been treated with pesticide. You are correct that the brown plant hopper is really a management problem. Under the research programme in Bengal, it was recommended that paddy transplanting be done in a north-south direction, which is the direction of the prevailing wind in Bengal during the paddy season. This ensured aeration of the crop and prevented build up of the brown plant hopper population.

Kenmore: I'm also concerned about the increase in pesticide use in India in paddy. Of the 400 odd districts that are significant rice-producing districts in India, 20 produce over 23% of the rice. If IPM is going to make an impact in India, as it has in Indonesia (Kenmore 1991), it must concentrate on the major producing districts; that's where syndromes caused by pesticide intensification will begin. If you apply the same argument to cotton, that's where you will find your pesticide resistance. Resistance isn't an issue in rice in India, yet. It has been in Taiwan and China for a long time. It has been in Japan for a long time. Increasingly, the immigrant brown plant hoppers that come from mainland China into Japan have been carrying a higher level of resistance, particularly to carbamates and phosphates. This resistance is due to an alteration in the acetylcholinesterase gene. Darwin is still just as right as he ever was—resistance will be a problem if you increase the selection pressure. You will get resurgence first, because you knock out the natural enemies. When you see the resurgence, that's where you should aim IPM to avoid later development of resistance.

The resurgence is your warning. If you manage that with the IPM training programmes in those areas, you will not get resistance problems. Once you have resistance problems, you can't go back.

Jones: Is this kind of story unique to paddy? Can we transfer the same ideas let's say to oversprayed cotton? Are there natural predators there? And is anybody doing this?

Kenmore: Yes.

Wightman: When a farmer changes from heavy insecticide use to use of no insecticide, in a 50 ha patch of paddy, for example, how quickly do the natural control processes take over?

Kenmore: One season sees a lot of recovery. After two seasons, the natural predator population in previously treated fields and untreated is not distinguishable. We talked yesterday about 500–600 species of natural enemies per hectare. You get close to that number between one and two seasons after stopping pesticide use on intensified irrigated rice. I agree it's one of the miracles on earth that this recovery is so quick. I think it's because there have been 50 000 generations of co-evolution among species in tropical rice.

Swaminathan: One important ingredient in changing pesticide use is the attitude of the politicians in government. As Dr Mehrotra said, in India pesticide use is increasing. Without public policy support, it is difficult to spread environmentally friendly methods of pest management. Workshops for policy makers on the social and public policy implications of IPM are therefore important. Are Indian policy-makers not as sensitive or receptive to these ideas as those in other countries in South-East Asia?

Kenmore: I have to tell the story about your successor as secretary, Mr S. Shastry. He came to Kanchipuram, Tamil Nadu, and had budgeted about half an hour for the visit to farmers in the field. He ended up spending three hours there. He thought he had been set up, because after an hour in the field, they marched him into a training centre where the DANIDA's *Women in Development* project had taken place. Women and men basically backed him up against the wall, and were asking questions. He had laid himself open to this by speaking Telefu with the people: there were enough people who could translate from Tamil to Telefu that they could go direct without a translator, so he didn't even have that buffer. He was standing in front of a picture of Krishna standing in the rice field with all the natural enemies coming out of his hand. The secretary was in a quite small room with 250 farmers—we are talking some serious political pressure. In the car on the way back, he agreed that all the pest surveillance centres and pest control centres, which are basically reservoirs of spray tanks, would be converted to IPM centres. That's a tactical answer—you set somebody up and you get them to the field, and you let the farmers convince them.

The big structural issue, in India and in China, is that these two countries have domestic pesticide industries. These are protected by the Ministry of Industry; they are not under the influence of the Ministry of Agriculture. The Ministry of Industry is trying to protect an indigenous pesticide industry and to take economic advantage, which means getting hold of the processing facilities for making products that are basically outdated and no longer used in other parts of the world. This is going to cause more and more trouble. You may get some pressure coming from citizens' groups. FAO can probably help put some pressure on generally to conform with what was said at the United Nations Conference on Environment Development in Rio de Janeiro 1992. Agenda 21, Chapter 14, includes a section on IPM.

Nagarajan: There has been much literature about the success of IPM for rice in Indonesia. India has avoided overuse of pesticide on rice and therefore has not reached the stage that Indonesia did a few years ago. The increase in the use of pesticide in rice is not as alarming as it has been in cotton. Available statistics indicate that about 14% of the pesticide used in agriculture in India is used for rice: 2% is fungicide, 3–4% is weedicide, and the remaining 9% is insecticide.

The other thing is that, unlike in Indonesia, in India there are varietal diversity and different growing seasons and many small farmers: these have all prevented pesticide abuse.

The Indian Council of Agricultural Research has taken up IPM in rice as a national priority; IPM for cotton and sugarcane will also be strengthened. So attempts are being made to contain the damage caused by pesticides before the situation gets out of hand.

In Indonesia, by withdrawing the insecticide in one go (which would be very difficult to do in India), they could suppress the population of brown plant hopper, which was the target pest at that particular time. It's only about 3–4 years since such an exercise has been done, and still there are other pests and diseases on the horizon. Are you confident that you will be able to combat, for example stem borer, blast, etc, which can come unforeseen and destabilize rice production? Or is there an alternate strategy being adapted to take care of aspects other than the management of brown plant hopper?

Kenmore: I agree that tactically the issue in India is to find the areas of intensified cultivation. The nationally aggregated statistics are not helpful, because India is not a country, it's a continent. There are many different parts of India, and one has to concentrate on the areas where the percentage of total pesticide use on rice is very high. There are districts in India where 90% of the pesticides used in that district happen to go on rice because that is the dominant crop.

We don't consider the brown plant hopper as a pest. We are not talking about a management strategy. A hopper infestation is an indicator that other things have gone wrong. The national policy declared by President Suharto in Indonesia was not pest control, it was a rice production policy. That's much more important than a pest control policy. Rice production is what keeps governments alive in Asia. If they can't control the price of rice at the appropriate level in the capital city, they will not stay in government in Asia. So it was a rice production policy and the brown plant hopper was the indicator that something had been going wrong. *What* was going wrong was shown to be overuse of insecticides because of another government policy.

Incidentally, if you want a cumulative figure, if you count a training day in the field as 5–6 hours, which is a long time to spend in the field, in Indonesia in the last three years we have organized more than two million person-days of training. This is probably the largest field environmental education programme ever attempted in the world, as well as being the biggest IPM programme ever attempted.

In that training, only 30% of the hours are spent talking about insects. They talk about agronomy, about crop physiology, and about sunlight. In Indonesia the long wet season is also the sunny season, which is different from the Philippines. The shorter drier season also has some cloud cover, so it's not as sunny. They talk about potential production, in terms that farmers can talk about. They know from their experience that they always get higher yields in the sunnier season. Why is that? Because the plants eat the sun, the insects eat the plant, the spiders eat the insects, the farmers eat the rice. Who eats the farmers? The money lenders! That's about as far as we can go in a government programme. The point is, it's not just a brown plant hopper management programme.

In 1989 there was a flare-up of stem borer in Indonesia. There was a move by people in the Ministry of Agriculture to release some of the pesticides banned under the policy for use against stem borer. That move was defeated by six other ministers: the Ministers of Development, Planning, Finance, Trade, Environment and Population and Health. The stem borer outbreak faded away. 300 000 people were involved in picking stem borer egg masses off of seedling rice—because it's all transplanted. It is a simple procedure to recognize the stem borer egg mass and take it off with your fingers when transplanting. In the post mortem of that outbreak, we looked at sub-district level records of the distribution of one of the pesticides that was still allowed under the new policy, namely carbofuran. We found a highly significant positive correlation between the tons of carbofuran distributed in the sub-districts and the subsequent outbreak of the white stem borer.

At that point we had a PhD student, Hermanu Triuidodo, from the University of Wisconsin, who is a native of East Java. He is a brilliant entomologist, and his team did the field trials. He looked at the survival of white stem borer larvae. He put first instar larvae on plants in fields. Some fields were untreated, some fields were treated with carbofuran. The survival rate in the untreated fields was 30%; in the treated fields it was over 60%. This means that you doubled the chance of survival by treating with carbofuran. So we think that we can handle other pests in Indonesia. I would also predict that in 3–4 years, Vietnam will have even a stronger programme of IPM than Indonesia does now.

Varma: In Kerala, frogs' legs were a major export for several years. The damage caused by brown plant hopper increased, because of lack of predation by the frogs, and the government banned further export of frogs' legs.

Insecticides have to be applied at the right time. The right time means at a particular level of the population of the insect. If insecticide is applied at the correct time, the insect population should fall. You are saying the pest population did not fall. Maybe the spraying was done too late. Do you have any information about this?

Kenmore: The issue is what's happening with the predators. When you say 'at the right time', you are just talking about the density of the pests and the life stage of the pest. Every time you use an insecticide, it's going to be killing predators as well as killing herbivores. So you have to look at the densities of both herbivores and predators. And you have to look at the potential population growth, which is a function of the age structure of the herbivores and of the predators. After you have done all of that, you have to start thinking about the nutrient qualities of the plant, whether it is a resistant variety or not, to give you another coefficient in front of that population growth rate. Then, on top of that, you have to start thinking about immigration rates, which come in later so you can't forecast them. I think the problem is soluble, once you have the instrumentation. I very much like Dr Jayaraj's suction machine that can collect populations to give you information on age structure. Dr Jayaraj was talking about capturing pests, but the machine can also be used for monitoring natural enemies. When you go to that level of detail, you can start making more intelligent projections about what will happen when you use an insecticide.

Varma: I agree, insecticides do not differentiate between pests and their predator insects. That's why in biological control experiments one looks from a different angle. Biological control does not depend on insecticides as such. In the Reunion Islands, biological control was done with no insecticide and it took two years to manage the psyllid population (Aubert & Quilici 1984). Biological control takes time. One does not use biological and chemical control together. But where insecticides are used as part of integrated management, if they are applied at the correct time and they are really effective, then they should reduce the pest population.

Neuenschwander: We are under pressure in Africa to switch from chemical to biological control in the same way. However, I don't think it will work in Africa as successfully as it has in Asia. We are doing training at all levels in about 30 countries; we have posters and calendars; we have two successful biological control programmes, and many more are being developed. But demonstration of successful biological control is more difficult than in the Asian rice systems for several reasons. First, in Africa we do not have the co-evolved systems of crops and pests; most of our crops are introduced. Second, we are talking about small areas of cultivation in a large area of uncultivated land, or land which at least is not cultivated at the same time. Thirdly, and probably most importantly, our systems are not as disturbed by previous use of insecticide as is the case in Asia. I work for an institute which caters to small-scale farmers; I'm not talking about cash crops. When an agroecosystem is not disturbed, it is difficult to demonstrate biological control, because the natural enemies are not numerous. There are a few pests, but those pests are parasitized to a very low degree; there is not much to show. So what remains to me is to do research on farmers' fields. I can document the impact of biological control on the pest, on the plant, on the yield and, finally, on the income of the farmer. Occasionally, although it is tricky, I can do an exclusion experiment: I can spray and get an explosion of the pest population.

The conclusion is that we are all in the same boat. We certainly want to learn from Peter Kenmore, especially about his teaching techniques. But, we have a different situation and Peter would be the first to acknowledge that the problems need indigenous solutions. In our work the farmers are not so directly involved, because we have relatively little to instruct the farmers, who do not use insecticides, what to do or not do. Our target is not the farmer, but national programmes and policy-makers.

Zadoks: I can quote you a screwed up system in Africa—groundnut and nematodes.

Neuenschwander: That is a cash crop.

Rabbinge: Cassava and mealy bug!

Neuenschwander: That's no longer screwed up (Herren & Neuenschwander 1991). It has been solved to a very large extent (95% of all fields) using biological control. I can demonstrate that it was biological control which was the key factor and not a new variety or the weather.

Jayaraj: What about spider mites on cassava?

Neuenschwander: IITA is working very strongly on that, and I think we now see the end of the tunnel.

References

Aubert B, Quilici S 1984 Biological control of the African and Asian citrus psyllids (Homoptera: Psylloidea) through eulophid and encyrtid parasites (Hymenoptera:Chalcidoidea) in Reunion Island. In: Proceedings of the 9th conference of the International Organization of Citrus Virologists. IOCV, Riverside, CA, p 100–108
David BV 1992 Pesticide industry in India. In: David BV (ed) Pest management and pesticides: Indian scenario. Namrutha Publications, Madras, p 225–250
Herren HR, Neuenschwander P 1991 Biological control of cassava pests in Africa. Annu Rev Entomol 36:257–283
Kenmore PE 1991 Indonesia's integrated pest management—a model for Asia. FAO Intercountry IPC Rice Programme, Manila

Communication and implementation of change in crop protection

M. M. Escalada* and K. L. Heong†

*Department of Development Communication, Visayas State College of Agriculture, Baybay, Leyte 6521-A, and †Entomology Division, International Rice Research Institute, Los Banos, Philippines

Abstract. The slow adoption of integrated pest management (IPM) has been attributed to the widespread gaps in farmers' knowledge of rational pest management. Other factors such as farmers' perception of high input use and promotion of pesticides also influence decisions to practise rational pest management. To bridge these gaps and improve farmers' pest management practices, most IPM implementation programmes rely on communication strategies. These communication approaches utilize either mass media or interpersonal channels or a combination. The choice of which communication approach to employ depends on project objectives and resources. Among extension and communication approaches used in crop protection, strategic extension campaigns, farmer field schools and farmer participatory research stand out in their ability to bring about significant changes in farmers' pest management practices. While extension campaigns have greater reach, farmer participation and experiential learning achieve more impact because learning effects are sustained. Communication media are important in raising awareness and creating a demand for IPM information but interpersonal channels and group methods such as the farmer field school and farmer participatory research are essential to accomplish the tasks of discovery and experiential learning of IPM skills.

1993 Crop protection and sustainable agriculture. Wiley, Chichester (Ciba Foundation Symposium 177) p 191–207

In many developing countries, farmers generally view the use of chemicals as the most effective and convenient method to control weeds, insect pests and diseases. However, farmers often use these at the wrong time and for the wrong type of problem, which can lead to resurgence of the pest and adverse environmental impact.

To improve pest management, farmers must learn field skills of pest and disease diagnosis, damage assessment, decision-making for chemical use, choice of crop varieties and optimal applications of fertilizer. They have to be motivated to analyse their situations, compare control options and choose that which best suits the situation. These principles have guided successful implementation of some integrated pest management (IPM) programmes, particularly those that consider farmers' perceptions and attitudes, informal networks, farmer participation, use of communication media and other socio-economic factors and use them as important influences in decision making. However, even well-designed schemes to implement IPM have often been slowed down by human and socio-economic constraints.

This paper presents an analysis of farmers' lack of knowledge and some influences in their decision making concerning pest management. We suggest approaches using communication to bring about changes in farmers' pest management practices.

Lack of information

There are gaps in current knowledge of farmers to make good decisions on the most suitable method of controlling pests. Results of farm surveys on farmers' perceptions and knowledge of pest management and current practice have often highlighted these gaps.

A survey of 300 farmers conducted in Leyte, Philippines, revealed a wide knowledge gap existing among farmers (Heong et al 1992). Although the main pests reported by the majority of the respondents were rice bugs, golden apple snails, leaf folders and lepidopterous larvae, in descending order of importance, most farmers believed the most damage was caused by tungro, a viral disease of rice. Yellowish to brown discolorations in rice plants may be caused by iron toxicity, organic soils, bacteria, insects (especially the white-backed plant hopper), fertilizer deficiency, water stress or phytotoxicity caused by herbicides. The symptoms may be distinguishable to the trained eye of the scientist, but farmers often attribute them all to tungro (Heong & Ho 1985).

The 300 farmers in the Leyte survey applied a total of 834 sprays; 37% and 25% of them were applied at the tillering and booting stages of the rice crop, respectively. The main targets were rice bugs, lepidopterous larvae and leaf folders. Farmers had grouped most lepidopterous larvae under the word *ulod* in Cebuano and Waray, which is a generic term for worms in these local languages. The pattern of insecticide use by farmers for various pests also raised questions on farmers' knowledge. Methyl parathion and monocrotophos were the main chemicals used against rice bugs, leaf folders and lepidopterous larvae, while Endosulfan was used for rice bugs, lepidopterous larvae and snails. Later, when farmers were asked which factors they considered in selecting the chemicals they used, they cited perceived effectiveness, price, endorsement from the extension technicians, availability and, to a large extent, advertising and promotion by chemical companies as important.

Another farm survey (Escalada 1985) revealed that while farmers were aware of and even planted resistant rice varieties, they did not know to which specific pests these varieties were resistant. IR-60, IR-42, IR-36 and IR-50 were the rice varieties most widely known among farmers. These are resistant to the brown plant hopper, blast and tungro. When asked what insect pests the varieties were resistant to, farmers frequently mentioned rice bugs, army worms, cut worms, plant hoppers and stem borers. Although nearly all rice farmers planted resistant rice varieties, more than half of them reported that they would spray these varieties as much as traditional ones.

The poor perception, knowledge and practices of farmers could be partly due to an ineffective transfer of information from sources to farmers (M. M. Escalada, unpublished paper, Int Ext Semin Exch Agric Ext Exper: Worldwide Trends Fut Needs China, 25–30 May 1992; Norton & Mumford 1982). One problem is that farming populations are not being reached by the agricultural extension service with the relevant information at the right time and frequency. Besides physical and logistical constraints, lack of transport and funds, semantic differences and psychological barriers also limit the access of farmers to information on pest management. Other problems include differences in the culture of the farmers and the extension technicians, and competing messages from multiple information sources.

Overestimation of the effects of highly visible pests

Farmers tend to over-react to visible pests or symptoms of damage and often overestimate yield losses associated with them. Working with Honduran farmers, Bentley & Andrews (1991) found that the importance of a farm problem was considered to be correlated with its ease of observation. Bentley (1989) described the farmers as knowledgeable about plants, which are generally large and stationary, but less so about insects, which are generally smaller. They know much less about plant diseases, which are usually caused by micro-organisms. According to Bentley, Honduran farmers also have a tendency to over-react to certain pests; for example, they spray insecticides on bean fields to kill *Diabrotica* spp. beetles even though the damage is minimal. The farmers perceive these beetles as threatening because they are large, active and brightly coloured and make large holes in the leaves of bean plants. English farmers also treat visible pests more readily but do treat invisible pests with preventive measures (Tait 1980).

In our Leyte survey, nearly all farmers thought that leaf-feeding insects would cause severe yield losses and would require spraying (Heong et al 1992). This belief may have accounted for the use of highly toxic insecticides, such as methyl parathion and monocrotophos, at the early crop stages. These chemicals are known to induce resurgence of the brown plant hopper (Heinrichs & Mochida 1984), thus these early sprayings would tend to put the crop at risk for secondary brown plant hopper problems at later stages (K. L. Heong, unpublished paper, Symp Migr Dispers Agric Insects, Tsukuba, 25–28 September 1991).

Risk aversion

Most farmers are averse to risk and seem to have little rational basis for use of pesticides; they tend to use pesticides for prevention. Our Leyte survey showed that rice farmers sprayed pesticides early, primarily for prevention or at the first signs of infestation with a pest or disease (Heong et al 1992). An earlier farm survey also revealed that 60% of farmers interviewed admitted they would spray insecticides even if they did not observe insect pests in their rice fields (Escalada 1985). They said this was to protect their rice crop from pests and to make sure that unseen pests would be killed. Because risk aversion has become deeply entrenched in many farmers, it is necessary to devise techniques and policies to wean farmers from their risk averse behaviour. Advertising messages often play important roles in increasing risk aversion. Having been made increasingly aware of pests by the chemical companies, farmers tend to over-react to slight symptoms. Without more objective information, the farmer becomes vulnerable to suggestions from these advertising messages.

Factors that influence decisions made by farmers

The association of high inputs with modernism

In the past, government-supported programmes for the intensification of rice production and advertising campaigns encouraging the use of pesticides have led farmers to associate pesticide use with modernism. A farmer who uses pesticides, even if the use is not warranted, is often perceived by his peers as up-to-date (Kenmore et al 1985, Bentley 1989). In surveys and focus group interviews conducted in the Philippines and Thailand, many farmers admitted that seeing their neighbours spraying their fields often prompts them to spray as well, even though it is not necessary (Escalada 1985, 1987). In China and Vietnam, farmers who do not spray are often classified as uncooperative and may even be subjected to certain sanctions by village authorities.

Promotion of pesticides

To increase sales, pesticide companies embark on massive advertising campaigns. Advertising messages tend to appeal to the emotions, particularly fear, of farmers rather than use rational arguments. In some countries, farmers are constantly exposed to colourful signs, magazine advertisements, stickers and television and radio broadcasts which remind them to use pesticides for production of a good crop. Sales representatives at community meetings and seminars often give away free samples, T-shirts, caps and other gifts. Because the primary objective of advertising is to increase sales, this leads to increased use of pesticides. One example is the excessive use of insecticides in Vietnam which seems to be associated with prominent advertising (Vo et al 1993).

The response of farmers in The Philippines to an extension campaign poster on IPM also demonstrates the extent to which pesticide advertising has become embedded in the consciousness of farmers. The poster was meant to promote IPM. It read: 'Be sure. Know the identity of rice pests. Choose the right pesticides. Don't let pests and diseases surprise you.' However, farmers misunderstood the intended message; the sprayer and the bottle in the poster led them to perceive that chemical control was being suggested (Escalada 1987, Heong 1989).

Ways of communicating information on crop protection

Widespread gaps in the knowledge of farmers and unfavourable attitudes of farmers toward rational methods of pest management impede the sustained adoption of IPM. Most IPM implementation programmes include extension or communication strategies which use the 'media' (e.g. radio, posters, pamphlets, stickers), direct personal contacts or a combination.

The radio campaign on IPM launched in The Philippines is an example of a media-based strategy (Pfuhl 1988). The campaign consisted of 48 radio spot announcements and 36 radio mini-dramas dealing with important issues of IPM in relation to rice. Both the announcements and mini-dramas were produced in two major Filipino languages, Tagalog and Cebuano.

Among the communication approaches that involve direct and mediated personal contacts, strategic extension campaigns, farmer field schools and farmer participatory research stand out in their ability to bring about significant changes in farmers' knowledge and practice of pest management.

Strategic extension campaigns

This methodology stresses the practical application of a strategic planning approach to agricultural extension and training (Adhikarya 1989). The campaigns focus on specific issues related to a given recommended agricultural technology. Their main goal is to solve or minimize the problems that prevent the technology being adopted by the intended target audience. The reasons farmers do not use a recommended technology, or use it inappropriately, are identified through a process that includes participation by the farmers. The results of a survey on farmers' knowledge, attitude and practice are used in planning campaign strategies and to develop messages for communication. The campaigns have to reach a large target audience rapidly and effectively. This is achieved most efficiently using a combination of mass media, group work and personal contacts.

Strategic extension campaigns require a multidisciplinary approach involving researchers, subject specialists, extension officers, trainers and communication support staff. These have to cooperate in the planning, implementation and management of the campaign.

The information derived from the survey forms a basis for monitoring the efficacy of the campaign. Two forms of evaluation are used: formative evaluation covers pretesting of materials and monitoring the management of the campaign; summative evaluation assesses information recall, i.e. how much of the information is retained by the farmers, and impact, i.e. the degree to which the recommended technology was implemented.

Integrated pest control—Philippines

A strategic extension campaign was carried out in rice-growing areas in Western Leyte, Philippines to raise awareness among rice farmers about IPM and to motivate them to ask for training. Communication was achieved through posters and audio cassette dramas that focused on the use of pesticides, when they were needed and natural enemies. There were also frequent field visits by two research assistants who took up residence in the campaign villages. As a result of the extension campaign, rice farmers in campaign areas drafted and sent petitions requesting the Department of Agriculture to provide them with training on IPM. In response to these requests, extension technicians in the campaign villages conducted a series of IPM training sessions for rice farmers. The results of the summative evaluation suggested that the communication campaign brought about significant changes in knowledge and attitudes toward IPM among farmers.

Golden snail control—Philippines

In the Philippines, the Department of Agriculture, the FAO and the International Rice Research Institute (IRRI) organized a strategic extension campaign to help farmers control the golden snail. Research-generated technology was not available, so baseline surveys and focus group interviews were conducted to learn about farmers' indigenous control measures. These were then tested for their effectiveness and modified to be shared with other farmers. The media used for the campaign were posters, stickers, slogans, comics and booklets, which were distributed directly to farm families or as training support materials in farmers' classes.

During the campaign, Department of Agriculture officers collected evidence on the toxic effects of chemicals being used by farmers to control the golden snail. The evidence led to a ban on the importation of the organotin compounds which were being marketed as molluscicides. An important feature of the campaign was the use of farmers' indigenous methods.

Weed management—Malaysia

To train farmers in chemical and other methods of weed control, the Muda Agricultural Development Authority (MADA) in collaboration with the FAO carried out a strategic extension campaign on integrated weed management. Launched in 1989, the campaign utilized media materials such as a campaign logo, dramas, songs, posters, leaflets, picture cards, flipcharts and a booklet. Summative evaluation results showed that 40% of the farm families adopted the recommendations and, specifically, could time their application of weed killer correctly. After the campaign, the Authority increased its total rice production to almost 10 000 tons, equivalent to US$2.33 million. The cost:benefit ratio of the campaign was 1:50 and the economic benefit per farm family that adopted the campaign recommendations was about US$191 per season (Mohamed & Khor 1990).

Media use in crop protection

These strategic extension campaigns concerning IPM demonstrated that different media could be used to improve farmers' pest management practices. However, the growing emphasis on media use has raised some doubts about its cost-effectiveness and impact in sustaining farmers' application of IPM (Bentley & Andrews 1991, S. Y. Chin, personal communication 1990). When mass media are regarded as sufficient means of communication and are used as substitutes for face-to-face interactions, field training and personal demonstrations of pest management techniques, then they lose their effectiveness.

Working on a pest management project in Honduras, Bentley & Andrews (1991) studied the impact of audiovisuals on extension. They found that while farmers and technical people liked the slide shows and pamphlets, there were no differences in the amounts farmers learned from various extension modes involving oral presentations. According to Bentley & Andrews, their study demonstrates that contrary to received wisdom, visual aids do not improve the effectiveness of efforts to extend complex technologies; they only add to the cost.

In the Philippines, surveys conducted to monitor the implementation of a strategic extension campaign on golden snail control revealed that a major constraint of the campaign was the over-reliance on media instead of live demonstrations in briefing farmers about the recommended control technology (Arguillas 1990, Dusaran & Canoso 1990). In some cases, the trainer did not even show a real specimen of a golden snail or demonstrate the mechanics of placing screen traps, canalets and stakes (Ramirez 1990). The trainers relied mainly on printed materials to teach farmers the methods for snail management.

Likewise, at a meeting to discuss the implementation of a campaign to control rats in Malaysia, organizers acknowledged that face-to-face training for extension workers was weak, that too much attention was paid to development and delivery of materials and not enough to 'preparing people to carry the message', and that extension workers felt uninvolved and unmotivated. While strategic extension campaigns can clearly bring about awareness and even improve knowledge in some cases, they cannot change practice. This is partly because media-based campaigns have little effect on changing attitudes toward risk aversion (Rogers 1983).

Farmer participation in IPM

Farmer field schools and farmer participatory research in the Philippines and Vietnam have demonstrated that farmers are enthusiastic in carrying out experiments and learning about IPM. Because such programmes can encourage farmers to make more rational decisions, their wider utilization should be encouraged.

As conventional extension systems are unable to reach the entire farming population, an attractive alternative is to encourage farmers to help other farmers. The effectiveness of this approach stems from the empathy, proximity and accountability of the farmers who provide information to their co-farmers. Previous studies had shown that the consciousness of the farmer tends to revolve around people close to himself, highlighting the central role of kinship and interpersonal networks of farmers (Escalada & Binongo 1989).

Mutual learning among farmers can be enhanced by tapping local networks, providing access to external resources whenever necessary and facilitating links with researchers and technical support. Basic investigation and validation of indigenous knowledge of farmers may be done by researchers.

Farmer field schools

In the field schools, farmers are taught to look at the rice paddy as an ecosystem. In Indonesia, 25 farmers meet for five hours once a week for 10 weeks. They observe plants and insects, then agree on how to manage the rice crop on the basis of their observations, experiences and weather forecasts. In addition to pest management, farmers learn about the application of fertilizer, water management and when pesticides should be applied (Stone 1992). The philosophy guiding the successful field schools in Indonesia is that farmers are very capable of training other farmers in a season-long process. Farmers graduating from the field schools receive certificates which allow them to train other farmers. They work in groups to pool skills and knowledge. Farmers work in the fields with support from the local IPM expert (Kenmore 1991).

Farmer participatory research

Farmers are encouraged to experiment to adapt new or existing technologies and spread them to other farmers (Bunch 1989). The advantages of experiential learning have been demonstrated in the diffusion of both traditional and recommended technologies in maize and cassava growing in West Africa, soil conservation techniques in Cebu, Philippines, and making contour ditches and planting Napier grass in Guatemala (Bunch 1989).

Simple rules for pest management, designed to enable farmers to experiment with them using participatory research, are being evaluated in the Philippines, Thailand and Vietnam. The aim is to convince farmers to withhold insecticide application in the first 30 days after planting. Avoiding early season spraying can reduce insecticide use of rice farmers by as much as 30%.

Each farmer measured off an area in the rice field where he or she did not apply insecticides in the first 30 days (the experimental plot). The rest of the field received the usual applications of insecticide (control). All other inputs and cultural practices were the same in both plots. The farmers also kept a simple record of their crop management practices over the entire rice season.

A primer on farmer participatory research was prepared as a handy reference to farmers during the conduct of the experiments. They were also provided with copies of the Cebuano version of the IRRI booklets, *Farmer's friends* and *Field problems of tropical rice*.

Over the entire cropping season, researchers visited the farmers regularly to establish rapport and monitor the progress of their rice crop, their pest problems and record-keeping. At the end of the season, farmers compared the yields of both plots. After the harvest, a workshop was organized where farmers shared their experiences with other farmers, as well as with extension technicians and researchers. Each farmer presented his or her results, the yields obtained from the control and experimental plots, and suggested an explanation for the differences in yield.

Thirteen of the 20 farmers had greater yields in the unsprayed plot with increases of 1–19%. Mean yields of the farmers were 149.3 cavans/ha[1] and 146.2 cavans/ha for the control and experimental plots, respectively. Seven farmers reported yield reductions ranging from 3%–18%. Explanations for the yield differences given by the farmers included differences in location of the two plots, loss of grains during and after harvest, differences in soil fertility, typhoon damage and pest attacks at later crop stages. None of the farmers attributed yield reductions in the experimental plots to not spraying in the first 30 days. They generally agreed that spraying during early season was not necessary. Fifteen of the 20 farmers said they would never spray during the early season in the future, while five said they would spray only if pest populations were high (Escalada & Heong 1993).

Farmers involved in participatory research are now into their second or third rice season and multiplier effects are evident. On their own initiative, the 20 farmers in the first group have taught 87 other farmers that early season spraying for leaf-feeding insects is unnecessary.

These experiences have shown that lateral learning can be effective and can potentially reduce the workload of scarce extension agents in Third World

---

[1] 1 cavan = 42 kg.

countries. Because farmers test an idea and validate scientific recommendations themselves, 'learning by doing' strategies appeal to farmers and the extension services in developing countries. They empower farmers with scientific knowledge which they can verify locally.

Conclusion

The power of communication in reaching and mobilizing individuals in developing countries has been widely documented. Well-designed educational and social marketing strategies can be utilized effectively to encourage more widespread and sustained adoption of recommendations by farming households. The choice of which communication approach to employ depends largely on the objectives and resources of the project. Because each strategy has its strong points and weaknesses, communication scientists have often argued for a combination of various media.

The strength of strategic extension campaigns lies in their ability to create awareness in farming communities of extension messages and agricultural recommendations relatively quickly and to mobilize farming populations into action. However, learning may not be sustained because mass communication is reactive and often the required interpersonal follow-up of face-to-face field training and on-farm demonstrations is not provided or continued.

In contrast, farmer field schools and farmer participatory research are interactive. These methods allow greater contact time with farmers, thereby encouraging a two-way exchange of information. Participation in a whole season's training or carrying out a simple experiment over one or more cropping seasons enhances farmers' self-esteem and gives them greater confidence in dealing with pest problems. Because of its experiential character, learning from participatory approaches tends to have more depth and is more sustainable. This observation is consistent with research in educational psychology which has shown that individuals retain 70% of the information taught if they learn it by seeing and doing (Minnick 1989).

A major disadvantage of farmer participatory approaches, however, is the relatively long process involved in achieving their intended learning effects because they can deal with only a few messages and farmers in a cropping season. While extension campaigns can be considered a macro-approach because of their wider reach, farmer participatory approaches tend to be limited in scope.

Given its complexity, IPM cannot be effectively communicated to farmers through media alone. Communication media are important in raising awareness and creating a demand for IPM information but interpersonal channels and group methods such as the farmer field school and farmer participatory research are essential to accomplish changes in attitudes and practice among farmers.

Acknowledgements

The authors wish to thank the Swiss Development Cooperation for funding the research reported in the paper through the Rice IPM Network based at IRRI, and FAO for supporting the strategic extension campaigns. Some of the ideas expressed in this paper were developed from these studies.

References

Adhikarya R 1989 Strategic extension campaigns: a case study of FAO's experiences. FAO, Rome

Arguillas J 1990 The management monitoring survey report on the strategic extension campaign on golden kuhol in Davao del Norte, Philippines. FAO Intercountry IPC Rice Programme, Manila

Bentley JW 1989 What farmers don't know can't help them: the strengths and weakness of indigenous technical knowledge in Honduras. Agric Hum Values 6:25–31

Bentley JW, Andrews KL 1991 Pests, peasants, and publications: anthropological and entomological views of an integrated pest management program for small-scale Honduran farmers. Hum Organ 50:113–122

Bunch R 1989 Encouraging farmers' experiments. In: Chambers R, Pacey A, Thrupp LA (eds) Farmer first: farmer innovation and agricultural research. Intermediate Technology Publications, London, p 55–61

Dusaran R, Canoso P 1990 Management monitoring survey report on the strategic extension campaign on golden kuhol in the province of Iloilo, Philippines. FAO Intercountry IPC Rice Programme, Manila

Escalada MM 1985 Baseline survey of rice farmers' knowledge, attitudes and practice of integrated pest control (Western Leyte): final report. FAO Intercountry IPC Rice Programme, Manila

Escalada MM 1987 Piloting a strategic communication campaign on integrated pest control in the Philippines. FAO Intercountry IPC Rice Programme, Manila

Escalada MM, Binongo S 1989 Diffusion and performance of indigenous and improved root crop varieties. Terminal report. Visayas State College of Agriculture, Leyte, Philippines

Escalada MM, Heong KL 1993 Human and social constraints to the implementation of IPM programmes. FAO Bull (15th Sess FAO/UNEP Panel Expert Integr Pest Control, 1992), in press

Heinrichs EA, Mochida O 1984 From secondary to major pest status: the case of insecticide-induced rice brown planthopper, *Nilaparvata lugens*, resurgence. Prot Ecol 7:201–218

Heong KL 1989 Sources of rice crop protection information. In: Harris KM, Scott PR (eds) Crop protection information: an international perspective. CAB International, Wallingford, p 51–66

Heong KL, Ho NK 1985 Farmers' perceptions of the rice tungro virus problem in the Muda irrigation scheme, Malaysia. In: Tait J, Napompeth B (eds) Management of pests and pesticides—farmers' perceptions and practices. Westview Press, Boulder, CO, p 165–174

Heong KL, Escalada MM, Lazaro AA 1992 Pest management practices of rice farmers in Leyte, Philippines. Survey report. International Rice Research Institute, Los Banos, Philippines

Kenmore PE 1991 Indonesia's integrated pest management—a model for Asia. FAO Intercountry IPC Rice Programme, Manila

Kenmore PE, Heong KL, Putter CA 1985 Political, social and perceptual aspects of integrated pest management programmes. In: Lee BS, Loke WH, Heong KL (eds) Integrated pest management in Malaysia. Malaysian Plant Protection Society, Kuala Lumpur, p 47–67

Minnick DR 1989 A guide to creating self learning materials. International Rice Research Institute, Los Banos

Mohamed R, Khor YL 1990 A summary of the process and evaluation results of the strategic extension campaign on integrated weed management in Muda irrigation scheme, Malaysia. FAO, Rome

Norton GA, Mumford JD 1982 Information gaps in pest management. In: Heong KL et al (eds) Proceedings of the international conference on plant protection in the tropics. Malaysian Plant Protection Society, Kuala Lumpur, p 589–598

Pfuhl EH 1988 Radio-based communication campaigns: a strategy for training farmers in IPM in the Philippines. In: Teng PS, Heong KL (eds) Pesticide management and integrated management in Southeast Asia. Consortium for International Crop Protection, Maryland, MD, p 251–255

Ramirez J 1990 Management monitoring survey report on the strategic extension campaign on golden kuhol in the province of Isabela, Philippines. FAO Intercountry IPC Rice Programme, Manila

Rogers EM 1983 Diffusion of innovations. Free Press, New York

Stone R 1992 Researchers score victory over pesticides and pests in Asia. Science 256:1272–1273

Tait EJ 1980 Environmental perception and pest control. Congress on future trends of integrated pest management. International Organization for Biological Control of Noxious Animals and Plants, Bellagio

Vo M, Thu Cuc NT, Hung NQ et al 1993 Farmers' perceptions of rice pest problems and management tactics used in Vietnam. Int Rice Res News 18:31

DISCUSSION

von Grebmer: We are dealing here with social change. I recently saw a nice slide which said: 'The only person who likes change is a wet baby'. We should not expect farmers to change their behaviour easily when we find it difficult ourselves. This also concerns industry. We have come from a situation where the farmers were told to use as much pesticide as possible to one where they are told to use as much pesticide as necessary. Industry has gone through a learning process, but this change within industry didn't come easily.

People change for three different reasons. One is incentives; one is punishment. The third is deeper insight or better insight. Most probably, we expect people to change too fast as a result of deeper insight. We should sometimes be a little more patient. If we expect too much too fast, we may encounter severe difficulties.

In marketing, at least among the soft science of economics, there's an old rule which says AIDA—Attention, Interest, Desire, Action. This is the sequence in which it normally flows. Do we have the attention of the people we want to involve? Have we aroused their interest or is it just a matter of courtesy that they are coming to the farmers' workshops? Do they have an actual desire to change? Finally, is there an action? I ask myself whether we use to the full degree what Kotler & Roberto (1989) have developed as social marketing tools. I find that in agriculture and agronomic policy we could use more of that type of social marketing, especially in developing countries.

Escalada: The AIDA principle has often guided past communication efforts to improve farmers' pest management practices. Although this principle seems to have been successfully exploited by the agrochemical companies, it has not met with the same success in IPM. Social marketing is best at creating awareness and promoting products among mass audiences, while participatory, experiential approaches work better with a complex philosophy like IPM.

In teaching IPM, we always start with rewards such as reduced expenses for chemical control and maintaining a better environment. We have learned that unless the farmers discover the mechanism behind these rewards themselves, learning will not be sustained.

Waibel: From what Dr Escalada has presented, one must conclude that it is the farmer who needs to change. But is this really true? In rice production, US $2 million are spent on pesticides annually. Take away Japan and it's still US $1 million per year spent on pesticides for rice. Then there is all the evidence which shows that the worst thing you can do, economically and ecologically, is put any insecticide on rice. 60% of trials over the last 10 years show no significant differences in yield. The results of trials that Dr Kenmore has conducted in Indonesia show that there is no relationship between yield and pesticide use per hectare. They show that as pesticide application per hectare is increased, the amount of rice produced per dollar of pesticide use falls.

Therefore, the problem cannot be solved by making the demand curve less elastic (i.e. such that if the price of pesticide falls, farmers won't use much more pesticide). The problem is to get rid of that demand altogether. One must ask: who really 'sprays'? It's not the farmer who sprays; it's also not the agrochemicals companies that spray; it is governments that make farmers spray because rice is a political crop. So is the target group really the farmer? I agree that training is terribly important, but not just for the sake of twisting that demand curve. Maybe the target group has to be changed.

Nagarajan: During on-farm training of cotton growers in Coimbatore district, printed material and coloured pictures are given out, also pheromones and light traps. The farmers have a choice and they select the items that they need; this is limited and is designed to act as an incentive to continue the adoption of IPM.

Wightman: One way to get more participatory research with farmers would be to close about 90% of the government and university research stations! Any agricultural scientists who survive this trauma should be supplied with a bicycle or, in the case of Vietnam, a boat. If they rode a bicycle and went and worked with farmers, they would have rather more time to think about what's going on around them. They could watch what's actually happening in fields and they could discuss with the farmers, on their own terms, what they should be doing. 90% of the work done in research stations has absolutely no relevance to what's happening in farmers' fields.

Kenmore: The issue is one of sustaining productivity and increasing productivity overall. We've talked about the yield plateau and the yield decline in intensified rice production. Dr Swaminathan initiated the study at IRRI and we have all been looking at that study very closely. One of the recommendations of Dr Pingali is that if farmers farm smarter, yields go up incrementally. You don't get big jumps, but there are incremental increases and there is lots of adaptation. That's shown by looking at differences among populations of farmers. Farmers who are doing better are innovating and are doing new things, and their yields go up. In the province around IRRI, the top one-third of the farmers are now routinely beating IRRI yields. This supports what John Wightman was saying.

It's not just a question of getting scientists out of the stations and down with the farmers. We also need to build up the tool box of the farmers. This is something that can be done. It can be done even though it requires more understanding—to think about managing a three-trophic level system instead of a two-trophic level system. It must be done because it is the way to generate new rice production.

Moni Escalada was talking about ownership: that's really the issue. If you discover something, you own it. In our field training courses, a farmer picks up an object and brings it to a trainer and says, 'what's this?' If the trainer says, 'that's a wolf spider' or 'that's a predator', the trainer is immediately fined. Answering a question with an answer kills learning. It reinforces the ego of the person answering, but it kills the learning process. All it proves is that the person answering is more powerful than the person who asked the question. If instead you say, 'It looks like a bug. Where did you find it?' The farmer says, 'I found it over there'. Then, ownership is beginning to transfer. 'Let's take a look. There are more of them, what are they doing? One of them looks like it's eating something else. What's it eating? The thing it's eating is trying to move.' Every step of that process, answering a question with another question, reflecting back, is a transfer of power, a transfer of ownership. It means the next day, or the next week, when the extension worker or trainer is not there in the field, a farmer doesn't have to remember what's a wolf spider and what is a brown plant hopper. The farmer remembers to ask simple questions and he can work out the knowledge. That's where you keep participation going. In that sense, we are working from both ends. We are working from where the farmers are, and from where the researchers are. The farmer participatory research Moni Escalada described is a step from the farming system style into a simplified and more widely applicable approach. We are working from the other side, from extension systems, building up farmer capability. The dialogue that's happening is on a much stronger level with a lot more sustainable concepts.

Klaus von Grebmer mentioned incentives and punishments and higher insight. Originally, ten years ago, we thought it was all incentive, but the punishment is there. If the farmers spray insecticide, they get an outbreak of brown plant hopper. They lose the rice and they are in debt for two years. We thought it was all going to be incentives. We said, 'you can make money'. The farmers replied that they don't look at the cost of pesticide as an investment; they look at it as a fear insurance. They think that if they don't spray, they will lose everything. The higher insight, about population change and about plant compensation, enables a farmer to look at this concept of a population threshold (as lousy as it is) in terms of an investment comparison, of how to invest resources, instead of as a fear insurance. The farmer needs to think that if he invests 25 000 rupiah now, he will get 40 000 rupiah later—to look at that cost as an investment. Then he can decide not to spend it on insecticide because that's not enough reward. He might want to spend that on a school book for his child or on hospital bills. He can spend it on something else, because he knows he's not going to lose the whole crop if he doesn't spend the money on insecticide. This comes from higher insight.

Saving money on insecticide is not a powerful incentive, because the amount of money involved is not big enough. It is much more important for the farmers to *understand* what the spiders are doing in the field. That turned out to be a much bigger selling point than I ever expected.

Ragunathan: I have been associated with organizing the training of extension personnel for 16 years. During such training sessions, the trainees admit that they invariably recommend application of pesticide without assessing the field situation. They realize later that most of their recommendations were faulty.

With the support of Peter Kenmore, FAO Regional Programme Coordinator, Inter-Country Programme for IPC in rice in South and South-East Asia, we organized two programmes for training subject specialists in Tamil Nadu and one each in Karnataka, Kerala, Orissa, West Bengal, Uttar Pradesh, Madhya Pradesh and Punjab. The extension workers were given intensive field training in the diagnosis of insect pests and diseases and weeds, as well as of the beneficial species that occur in the rice ecosystem. Because most of the extension workers were not familiar with the beneficial species and their role in the natural suppression of pests, most of the training was done in the rice fields. In Tamil Nadu, 50 subject specialists trained under this programme then passed the training on to 700 extension workers within two years through similar programmes.

Realizing the benefits of IPM, the Department of Agriculture conducted a series of IPM trials at different locations. On average, each farmer had a net gain of Rs. 1000 per acre. So far, the State Government of Tamil Nadu has organized IPM demonstrations on more than 40 000 ha. Currently, these efforts have resulted in a saving of more than Rs. 30 million in pesticide subsidy.

In Andhra Pradesh, the brown plant hopper has been tackled effectively in coastal districts by avoiding pesticide spraying. Now there are no outbreaks of the brown plant hopper in these coastal areas. As a result, the State Government has saved a lot of money on pesticides. These are some examples of success stories of IPM in India.

Upton: I would like to comment on people's attitudes and objectives. There is very often a big difference between what people say their attitudes and objectives are, and the way they actually behave. We should base our observations on what people do rather than on what they say they do. One example comes from a Ciba Foundation debate on organic farming. A supermarket chain in the UK studied consumer attitudes to organic foods. They asked consumers whether they thought organically produced foods were healthier than non-organic foods. 80–90% of respondents thought they were healthier. 40–50% said they would be willing to pay more for organically produced foods. But when the supermarket chain checked the records of their sales of organic foods, which are labelled as organic foods, only 10% of purchases were organic foods. This illustrates the point that people behave differently from what they say, but it may also give an indication as to whether consumers would be willing to pay more for food produced by IPM methods.

The other illustration is the argument used by many economists. We know that people have many objectives, but farmers and consumers and others behave *as if* they are influenced by profits or by money values. There is a great deal of evidence from all over the world, including West Africa, that farmers are responsive to prices. They are driven by economic incentives. So if I earlier said there was no competition between food and cash crops, I would like to correct that. When prices of cash crops are higher, when cash crops are more profitable than food crops, farmers do switch to cash crops. This suggests that economic incentives are important in the adoption of IPM. I'm surprised that farmers continue to use insecticides, if they are as damaging to farmers incomes as we've been told this afternoon.

In my paper (Upton, this volume), I suggested that there were areas where farmers are not aware of all the costs and benefits or don't experience all the costs and benefits of their actions. There are externalities, for instance, in the use of insecticides. I said these situations require government intervention, but this assumes benevolent and well-informed governments. We have been given instances of the opposite—of government failure, in contrast to market failure. I don't know the solution to this, except for people like Peter Kenmore and Moni Escalada, who are well informed and well motivated, to continue influencing events. One can't rely simply on farmers' knowledge and farmer participation to lead to the right policies. There are external costs and external benefits of which the farmers may not be aware.

References

Kotler P, Roberto EL 1989 Social marketing: strategies for changing public behavior. Free Press, New York

Upton M 1993 The economics of food production. In: Crop protection and sustainable agriculture. Wiley, Chichester (Ciba Found Symp 177) p 61–75

Plant diseases in India and their control

S Nagarajan

Indian Council of Agricultural Research, Krishi Bhawan, Dr Rajendra Prasad Road, New Delhi 110001, India

Abstract. The concept of development is reviewed in terms of sustainability. Food production in India driven by pressure from an increasing human population uses 90 000 t per year of technical-grade pesticide: 12% of this is fungicide and a good part is insecticide for the control of vectors of plant viruses. A change in the cropping pattern and irrigation have provided a summer 'green bridge' along Tamil Nadu/Andhra Pradesh border areas for the tungro virus that affects rice and its vector. Epidemics occur along the coramandal coast, if the weather is suitable. Red rot disease of sugarcane is promoted by poor drainage, river widening, ratooning, contaminated planting material and variation in the pathogen throughout the Indo-Gangetic plain. Apple production uses large amounts of fungicide. For every 1000 t of apples produced 1t of fungicide is sprayed 8–10 times seqentially. Systemic application of fungicides has led to pesticide resistance and resurgence of other diseases. 70–80% of the Nagpur Mandrin produce reaches the market by trucks that have to traverse 1000 km. 10.6% of fruits are lost to post-harvest diseases; culling, sunburn and injuries account for another 11.6%. In the control of leaf rust of wheat in North India, the use of varietal mosaics, resistance genes and extra-late wheat sowings that do not coincide with favourable weather have all collectively contributed to loss reduction. The drop in the production of exportable crops such as peppers and coconuts because of diseases needs attention. The traditional wisdom on crop mixtures, organic manuring, shifting sowings, etc, needs scientific re-evaluation.

1993 Crop protection and sustainable agriculture. Wiley, Chichester (Ciba Foundation Symposium 177) p 208–227

Indian agriculture is as old as the human ingenuity of cultivating plants. Spectacular strides in productivity have been made during the last three decades owing to increased use of fertilizers, pesticides and water. The rise in food production has resulted in higher availability of food per capita, and has also met the additional demands created by better living standards (Table 1). The agricultural economy registered an upward trend during 1986–1990, thus even during the era after the green revolution there is sustained growth. The value of agricultural output went up by 26.6% and in 1992 was worth Rs 620 thousand million (1US$ = Rs 29.5). Of this increase, better crop husbandry accounts for 23.6% and the animal husbandry sector for 19%. Such continued growth of

TABLE 1   The human population in India, the area and production of cereals and annual income per capita

Year	Population (million)	Area (Mha)	Production (Mt)	Income (Rs)
1975	62.0	101.4	128.1	1464
1980	68.9	104.5	144.8	1627
1985	76.9	105.4	164.5	1813
1988	81.9	102.6	175.6	2082

the agricultural sector over the last three decades and industrial development have created problems with regards to the health of the soil, plants and the environment.

The changed agricultural scenario has focused attention on ways to check plant diseases, cut crop losses, avoid wide fluctuations in production and sustain the higher levels of productivity. Application of pesticides was introduced in the mid-1950s: in 1992 sales of pesticides were around Rs 8000 million, sales of seed were Rs 6500 million and those of plant protection machinery and appliances another Rs 1000 million. This plant protection market is still growing and changing simultaneously.

The multiplicity of crops, overlapping growing seasons, different varieties and cultural practices favour the occurrence of an enormous number of plant diseases. All of them impair production; some of the most important diseases are listed in Table 2. In this brief paper I shall not attempt to cover all of them, but will try to project the complex and dynamic state of the situation and the growing demand for efficient management strategies.

Rice tungro disease

In India, 41.2 million ha of rice (*Oryza sativa* Sub.sp.*indica*) are grown in diverse environments and cropping sequences. Of this, 28 Mha are planted with the high-yielding varieties. The paddy production is 110.5 Mt, which when milled comes to nearly 80 Mt of rice. 12 000 Mt of technical-grade pesticides are used to control insect pests, vectors, plant diseases, rodents and weeds. Leaf yellowing in rice is transmitted by a vector that was shown to be a virus (Raychaudhuri et al 1967), later identified as rice tungro virus (John 1968). From the northern states where it was first noticed, the virus moved south and has there become endemic. The infected plants are stunted, pale, shy tillering, occasionally with distorted leaves. Tender leaves may show mottling and the older ones become rusty in colour. Plants infected early die, but others survive and produce a weak earhead that seldom gets filled.

TABLE 2 Major crop diseases important in Indian agriculture

Crop	Pathogen
Rice	*Pyricularia oryzae, Xanthomonas campestris* var. *oryzae,* tungro virus
Wheat	*Puccinia recondita tritici, Ustilago nuda*
Sorghum	*Spacelotheca sorghi, Colletotrichum graminicola*
Maize	*Setosphaeria turcica, Diplodia maydis*
Pearl millet	*Sclerospora graminicola, Claviceps fusiformis*
Sunflower	*Alternaria helianthi, Puccinia helianthi*
Ground nut	Tomato spotted wilt virus, *Puccinia arachnidis/Mycospharella arachnidis*
Sugarcane	*Glomerella tucumanensis, Ustilago scitaminae*
Tobacco	Viruses, *Meloidogyne* spp.
Soybean	*Macrophomina* spp., bud rot
Chickpea	*Fusarium* spp., *Macrophomina* spp.
Pigeonpea	*Fusarium udum*, sterility mosaic (?)
Banana	Bunchy top virus, *Fusarium oxysporum*, f. sp. *cubense*
Apple	*Venturia inaequalis*
Citrus	Die-back syndrome
Grapes	Mildews, root knot nematode
Mango	Malformation, *Oidium mangiferae*
Potato	*Phytophthora infestans*, viral diseases
Vegetables	Vein clearing disease (Bhendi), *Heterodera* spp.
Tea	*Exobasidium vexans*, root diseases (*Fomes, Rosellinia*, etc)
Coffee	*Hemelia vestatrix*
Rubber	*Phytophthora palmivora*
Ginger	*Pythium aphanidermatum, Fusarium* spp.
Spices	*Phytophthora palmivora* (pepper), soft rots of cardamom
Coconut	Root wilt, *Phytophthora palmivora*

In part from Geddes & Iles (1991).

During 1990, 55 000–60 000 ha in the west Godavari district were affected by tungro, of which 2000 ha were totally lost. Many thousands of hectares along the northern arm of the tungro path (Fig. 1) were badly damaged. Several years before this, there was an epidemic on the southern side. In 1992, by July, 2000 ha were damaged by tungro in the northern districts of Tamil Nadu, though the usual season for outbreak is November. During the 1980s the disease became endemic in the delta areas of the eastern coramandal coast, between Vijayanagaram/Puri in the north and Pudukkotai in the south. In this tract, there has been no substantial change in area and rice cannot be grown without abundant water. But the continuous overlapping of the rice crop creates a

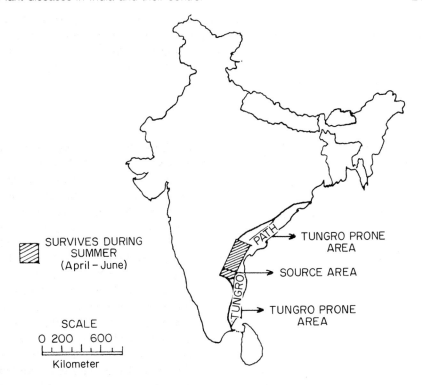

FIG. 1.   The postulated source area and the tungro path over south and peninsular India.

'green bridge' on which the vector and the virus survive, even during the summer months.

Nature of the disease

Tungro exists as two associated viral particles: a rod-shaped form ($35 \times 175 \ \mu$m) and a spherical form ($28 \ \mu$m). Both have to be present in the host for expression of the symptoms. These particles are differentially transmitted by the rice pest called the green leaf hopper. The two species of hopper involved are *Nephotettix virescens* and *N. nigropictus*. The vector needs to feed on the infected plant for at least 10 minutes to acquire the viral particles; after that, it is able to transmit them successfully in a non-persistent manner.

Many species of *Oryza* are susceptible to tungro, though expression of the symptoms varies considerably. Apart from the self-sown and ratoon crop of rice, the weeds that commonly occur in the rice ecosystem act as collateral hosts for the vector and the pathogen. The occurrence of tungro throughout the year in Tamil Nadu and West Bengal, where rice is grown continuously in overlapping crops, permits the survival of tungro complex on the main rice crop itself. *N. virescens* is the most efficient vector and prefers rice, while the less efficient *N. nigropictus* prefers grasses (Rao & Anjaneyulu 1978, Anjaneyulu 1986). The pattern of epidemics each season differs depending upon which vector species and which biotype of the virus predominates.

Vector dynamics

The All-India Coordinated Research Project on rice at Hyderabad has monitored the population of the green leaf hopper at various locations in India during the rice growing season using multilocational light traps. Two years' worth of their unpublished data were analysed. Unfortunately, information on the percentage of hoppers that are viruliferous, the vector species involved, the viral particle(s) that they carry, the extent of tungro disease during different seasons, etc, is not available, which makes critical epidemiological analysis difficult. Yet the insect catches made at five locations during the off-season of March–May 1989 had a maximum cumulative catch of 85 000 green leaf hoppers at Warangal, Andhra Pradesh (Fig. 2). Locations in northern India recorded much smaller catches. The southern belt invariably had a high population of green leaf hoppers. The main rice-growing period is May to December. Warangal recorded nearly 200 000 green leaf hoppers between mid-August and the end of November. At Maruteru the catches were comparable to those in the off-season (Fig. 2). In 1990, the year of the epidemic, for the corresponding period Warangal and Maruteru recorded a catch which was more than double the total trapped population for 1989. The belt around Warangal seems to be the area in which the hopper survives. The tungro-resistant varieties of rice in the area are maintaining resistance, therefore the epidemic was not due to any new strain of either the virus or the vector. Abundance of the vectors seems to be the main reason for the epidemic.

The tungro path in the south

In northern India (excluding Bengal and the eastern states), only one crop of rice is grown and hence the vector population dwindles during the winter. In the south, vector populations were very small at Aduthurai and Coimbatore in Tamil Nadu during January to May (Fig. 2). However, at both Maruteru and Warangal during March/April thousands of green leaf hoppers were trapped, which clearly indicates that a green bridge is operative in the area. The summer rice crop grown in southern Andhra Pradesh and northern Tamil Nadu assists in the perpetuation of the disease. From this source area (Fig. 1), the vectors move along the eastern coast, possibly under the influence of the July to September monsoon winds, which also bring the rain needed for rice transplantation. These new epidemiological insights can be used to design an effective strategy for disease management.

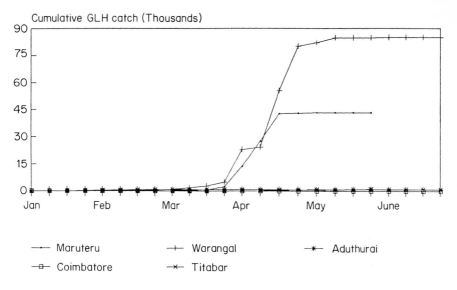

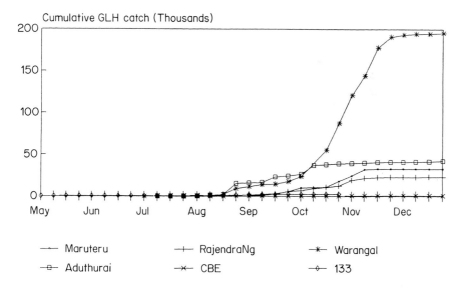

FIG. 2.   The cumulative weekly catches of the green leaf hopper (GLH) from light traps
in several rice-growing locations during 1989. Top: January to June, including the off-
season of March to May. Bottom: The main rice-growing season of May to December.
Note the scales of the vertical axes are different. Data from unpublished annual report
of the Project Directorate (Rice), Hyderabad.

Effect of chemical control of tungro

Several insecticides, such as phorate, diazinon, carbofuran, fensulphothion and cypermethrin, are effective against the vector. Cypermethrin quickly kills the vector even at low dosages. Application of insecticides such as carbamyl has increased over the years in the source area of the vector (Table 4). Indiscriminate usage of insecticides and cultivation of susceptible varieties of rice like IR-64 fuel the tungro epidemic.

In the rice fields there are 10 species of spiders belonging to seven families, of which five are webspinners. There is a large population of predators, parasites and beneficial fungi, nematodes and amphiba in the rice ecosystem (Table 3) (Bandyopadhya et al 1986), which keeps the population of green leaf hoppers under control. When a super-susceptible variety like IR-64 is cultivated or the agronomy is changed, the pest population explodes. This is countered with a pesticide spray which further damages the rice ecosystem, creating a demand for more and more pesticide. The insecticides recommended to check the green leaf hopper also act against the biocontrol agents and so the hopper population builds up much faster after the spray, resulting in the need for additional rounds of spray. Now rice is following the course of cotton—getting trapped in the pesticide treadmill.

Rice tungro management

It is now obvious that the vector spreads primarily from a small area where it survives during the summer. It is therefore advisable that varieties with different resistance genes are released along the tungro path to reduce the vector population (Fig. 1). The success of the programme to contain brown rusts in wheat purely by varietal deployment shows that such an approach is feasible (Nagarajan & Joshi 1985). Continuous monitoring of the vector population during the summer and serotyping to estimate the proportion that is viruliferous will assist in making decisions on the nature of control actions. The trap counts of the green leaf hopper and weather-based population forecasts can be used to predict the anticipated hopper population, enabling timely action to be taken. Enriching the biocontrol agents in the rice field, adoption of varieties like Vikramarya or Srinivas and suspension of all unwanted insecticide sprays will lead to more effective control of tungro.

Red rot of sugarcane

Sugarcane is one of the major commercial crops in India and meets the demands of more than 400 sugar mills. Over the years there has been a steep increase in sugar output since the canes for Khandasari (non-clarified sugar) were diverted to the mills. Of the various pests, *Pyrilla* and red rot are very important.

TABLE 3   Natural enemies of pests noted in a rice ecosystem in West Bengal that also feed on the tungro vector, the green leaf hopper (Bandyopadhya et al 1986)

Parasites of egg	Parasites of adult	Predators of insects	Predators of spiders
Gonatocerus	Helictophagus	Cyrtorhinus	Paradosa
Mymar	Hexamermis[a]	Ophionea	Oxopes
Anagrus	Beauvaria[b]	Paederus	Zygoballus
Oligosita		Crocothemis	Thomisus
		Agriocnemis	T. cherapunjiensis
		Microvelia	Tetragnatha
		Conocephalus	Clubiona
		Micraspis	Neoscona
		Nabis	N. nautica
			Crytophora

[a]Fungus.
[b]Nematode.

The insect *Pyrilla* has been very effectively biologically controlled over the Indo-Gangetic plain, but red rot continues to be serious. Several cane varieties have been released for this tract but they have a field life of barely 7–10 years and succumb to the disease.

The annual recurrence of red rot disease is through infected sett, stubble and propagules that may survive in the soil. From the primary infection centre that is established, the disease spreads up to 6m and causes drying of the infected canes (Singh 1988). The conidia spread through irrigation water and infection occurs through the nodal region. Conidial fusion is a regular and normal feature, which contributes to the creation of variability in the pathogen and pathogenic forms that result in varietal breakdown (Duttamajumder et al 1990). More quantitative studies to validate the pathogenic forms and the basis of disease resistance are in progress. Disease is therefore being managed through the use of healthy planting material, proper field levelling and drainage, sanitation, a reduction in the number of ratoons and by the use of resistant varieties.

Coconut root (wilt)

Of all the coconut-producing states, Kerala has the largest area under cultivation and the lowest production of nuts per unit area. While the Tanjore wilt (*Ganoderma lucidum*) is important in Tamil Nadu, the controversial root (wilt) disease is the major constraint in Kerala. The disease is found in a contiguous tract stretching from Trivandrum to Trichur: in the Allepy, Quilon and Kottayam districts 50% of the trees are diseased (Fig. 3). Circumstantial evidence favours a mycoplasmal aetiology of the early symptoms, namely leaf flaccidity, shortening of the whorl, a smaller crown and fewer nuts (Solomon et al 1983),

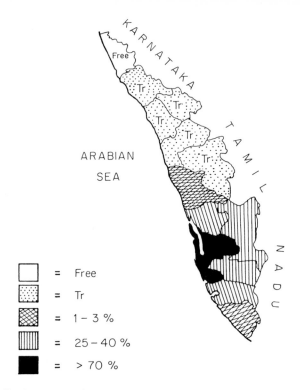

FIG. 3. Distribution map of coconut root (wilt) disease in Kerala, India. From Jayansankar & Bavappa (1986).

which predispose the tree to the killer leaf rot (*Bipolaris*) disease. The banana lacewing bug (*Stephanitis typica*) is branded as the vector, though insects in this group are not the usual vectors of mycoplasmal pathogens. In the diseased palms, the crown is shortened, the phloem is choked and the permeability of the root system is altered (Jayasankar & Bavappa 1986). Secondary infestation of these indisposed trees by an array of organisms, like nematodes, pathogenic bacteria and the *Bipolaris* that causes leaf rot, results in further weakening and even death of the tree. In plantation crops that have a long gestation period, breeding for resistance and replacement of varieties is very tedious. Because diseases that cause decline are due to complex factors, isolation of suspected trees, sanitation through removal of infected trees, proper nutrition and good agronomic practices are the only means of disease control apart from the possibility of checking the vector.

Post-harvest diseases

Fruits and vegetables suffer incredible losses due to diseases, even when they stand in the field and later after harvest when they are transported and marketed.

Because fresh fruits are preferred to canned ones, loss due to post-harvest damage is very high. The extent of and reasons for damage after harvesting have been investigated for the mandarin oranges that are grown around Nagpur, Maharashtra and marketed all over India (S. A. M. H. Naqvi, personal communication). Loss from the time the fruit is picked until it reaches the local wholesaler is 8.4% and another 20% is lost when the fruit is transported by truck to Delhi. In 60 hours a truck transports 600 boxes of fruit to Delhi, which is 1200 km north, while a rail wagon carrying 750 boxes takes 70 hours. If transported by rail, losses are 22%. Further losses occur between the Delhi market and the consumer. The reason for these high losses is that the fruit is handled roughly at all stages at least twenty times before it reaches the Delhi market and many more times thereafter, until it reaches the customer. Bruises, sunburn, stem end rot and sour rot are the major types of post-harvest damage. By following simple prophylactic procedures, growers and traders could reduce the losses, but such measures are not being taken. At present, the consumer is paying for the mistakes and negligence of other people. Only the pressure of a competitive market for produce at a price affordable by the consumer will promote the elimination of avoidable losses.

Approaches to disease management

Overuse of pesticides has caused serious problems. Ideas on 'integrated' and 'sustainable' agriculture have captivated the general public. The implications of pesticide abuse have gone through several phases of emotional, technical and economic questions. Politicians and governments are under pressure to initiate change (Zadoks 1992), therefore approaches to the control of plant diseases will be discussed in view of the present demand for a cleaner environment.

Effects of agronomic changes on plant diseases

Intensive cropping in Punjab over the last 25 years has resulted in the depletion of soil organic matter. The carbon content has fallen from 1.2% to 0.3% (Sidhu & Byerlee 1991). This has probably contributed to a reduction in the population of saprophytic fungi and flora of the rhizosphere, resulting in a gradual decline in the yield of the rice/wheat rotation system. Along the foothills of north-west India, traditional farming practice before the mid-1960s was to grow chickpea and wheat together as companion crops. The good price for wheat and high-yielding wheat varieties led to pure cropping of wheat and infestation with Karnal bunt (*Neovossia indica*) became serious. Two years of experiments at Dhaulakuan in the Himalayan foothills, using various mulch and companion crops like chickpea or mustard along with plastic mulching, showed that chickpea reduces the Karnal bunt by a maximum of 57% (Table 4) (Singh et al 1991). Because the primary inoculum, chlamydospores, is soil borne, mulching the soil

TABLE 4   Effects of mulching on Karnal bunt incidence at Dhaulakuan (Himachal Pradesh) under uninoculated conditions

	1988		1989	
Treatment	A	B	A	B
Wheat + mustard	0.182	9.4	0.080	38.16
Wheat + chickpea	0.122	39.3	0.057	57.25
Wheat + plastic	0.075	62.6	0.033	74.8
Wheat (check)	0.201	—	0.313	—

A, average % infection; B, % reduction of infection.

creates a physical barrier between the source of the pathogen and the emerging earhead which is the site of infection. Native intelligence made the farmer realize this and plant the correct companion crops. The cropping procedures and sequences are changing rapidly and it is now felt desirable to record these good agronomic practices so that future researchers can draw upon this information to develop better sustainable techniques.

Not all the changes are bad. Over north-west India, rice was restricted to a small area and was cultivated between June and November. During the 1980s, India emerged as a rice exporter mainly owing to an unbelievable increase in the area planted with rice in this tract. The good soil, irrigation, climate, quality of rice and the price made this rapid expansion in rice cultivation possible. As inputs are not always available in time, and the quality of rice is better if it matures during a cooler period, wheat sowings were gradually postponed. The less attractive price for wheat made the farmers slowly switch their wheat sowings, which used to be in late October, to early January. Other crop sequences, such as an autumn crop of potato, also became popular. Consequently, now nearly 30% of the area under wheat is sown after mid-December and 15% of the area is devoted to super-late January sowings.

This changed agronomy has markedly altered the epidemiology of both *Puccinia recondita tritici* and *P. striiformis striiformis*. For the last 15 years, even though matching virulences have occurred in the hills, severe epidemics have not occurred. Now, when the primary inoculum arrives from the hills during mid/late-December, a large area is still unsown and so the initial level of inoculum (Vanderplank 1963) of the disease gets reduced. The weather conditions for the matching growth phase for the super-late sowings are too cool for the disease to become established. This sanitation effect has imposed a time delay on the epidemic. The dry weather in April and the accelerated maturity due to the desert winds do not favour rust development after mid-April. Coupled with this, the resistance of the adult plants, varietal deployment, etc, have also contributed to the delaying of the epidemic and the low terminal disease severity. Hence, a change in agronomy contributed to a 10% increase in yield which made even super-late sowing of wheat economical.

Disease resistance

Since the days of Biffin, breeding for disease resistance has been one of the most favoured approaches in disease control. Advances in procedures for storing inoculum, mass seedling inoculation and in-field epiphytology, as well as the development of handy scales for scoring disease severity and physiological growth stage, etc, have accelerated the pace of breeding high-yielding disease-resistant varieties. This programme, primarily run by the state, is gradually being taken up by the private seed companies and this may have some implications for pest management strategy (Nagarajan 1992).

In the progressive states of north-west and peninsular India, the average lifespan of cereal varieties is about five years from the time they are passed to the farmer for cultivation. In other states, such as Bihar, it is as high as 12 or more years. In these states the seed replacement ratio or the percentage of farmers who purchase fresh lots of seeds is very low; once a farmer gets seeds of a particular genotype, he changes reluctantly if at all. But in the progressive states, once a genotype becomes susceptible, it is replaced by a resistant one, or when there is a better variety it is quickly adopted. In states like Punjab, a predominant variety can be totally withdrawn and replaced by another within three years. This speedy varietal reshuffle has been one of the cardinal principles for the management of wheat rusts in the northern states. The success of varietal reshuffle should be credited to the All-India Coordinated Project, which through its vast network of multilocation evaluation was able to distribute new varieties. It is in this context that intellectual property rights and other issues are important: the national pest control strategy should be reviewed periodically, taking into account new developments and the changing scientific climate.

There is virtually no arable crop in India that has not been genetically improved for pest resistance. This labour-intensive pest management procedure is suited to a country like India, where most of the crops can be grown at least twice in a calendar year, where trained labour is abundant and where there is no restriction on access to germplasm. The national plant disease programme should continue to derive its strength from this for some time to come and by giving equal partnership rights to the plant pathologists, entomologists and agronomists much more benefit can be derived from the All-India Coordinated Research Project.

Disease prediction system

The availability of computers has totally revolutionized data management. Close monitoring of the progress of epidemics and weather conditions has led to the development of linear decision-making models which have been integrated as a simulator. The rice blast simulator EPIBLA (Rao & Krishnan 1991) is an exercise that needs further field validation. The classical success has been the apple scab prediction procedure that was developed and validated for field work in the states of Himachal Pradesh, Jammu and Kashmir. Through this programme it has become possible to rationalize fungicide spraying for the control of apple scab disease.

Production of healthy seeds

Many plant pathogens survive latently inside seeds and other propagules and the disease expresses itself the following season when the infected materials are sown or planted in the field. Seeds are dressed with several fungicides to prevent infection. In addition, appropriate pre-harvest procedures have been developed to produce good quality, healthy seed to meet the quality standards. Unfortunately, there is no legal sanction to prohibit the sale of infected seeds: as a result, the clandestine trade in infected and sub-standard seed is flourishing. Potato is the classic example of an increase in productivity through the use of quality seed. Techniques for producing seed tubers free from major viral diseases include monitoring the aphid vector population, spraying insecticide only according to need, sero-indexing of healthy tubers and earlier cutting of the haulms. These simple techniques were developed by the Central Potato Research Institute during the 1970s for the production of cheap, disease-free seed tubers. This technology assisted in increasing potato production by 600% in about 20 years.

In Punjab, loose smut of wheat continues to cause 2% yield loss. The earcockle of wheat still occurs in north-west India, and the mung bean (*Phaseolus mungo*) mosaic virus, which is seed borne, is a major production constraint in all the pulse-growing areas. All these testify that seed health remains a neglected activity. Through the use of contaminated seed, plant pathogens can be transmitted across the country, adding to the problem of disease control. The productivity of horticultural crops, particularly of fruits and flowers, can be substantially increased by certifying or registering the nursery owners and introducing a common code of practice.

Rationalizing pesticide usage

In India nearly 90 000 Mt of technical-grade pesticides are used per year. 12% of the fungicides and many of the insecticides are used to control the vectors of plant viruses. On plantation crops, fruits and vegetables maximal amounts of fungicides are used to produce commodities of high quality that fetch a good price in the market. Overusage of sterol-inhibiting systemic fungicides induces fungicide resistance in the pathogen and the selective fungicides trigger resurgence of other minor diseases. Various strategies have been suggested for the optimal and need-based use of these chemicals, so that their beneficial agricultural span can be increased. Screening for new molecules that have fungicidal properties and developing them as a marketable formula is both capital and technology intensive. The cost structure of launching a new product of uncertain market makes it compelling to adopt an integrated system of management rather than being dependent on one approach for solving particular problems. Since once-dependable products are being withdrawn or reviewed because of health reasons, it has become necessary to combine relevant minimum plant protection steps as an integrated pest management (IPM) package and to promote their adoption.

Total withdrawal of fungicides and replacement of the control strategy with biological and other means will be difficult in practice. Therefore, withdrawal of leading fungicides will only result in increased crop loss. The GRC economic survey conducted in the USA during November 1989 indicated that in such a situation prices of many commodities would increase substantially, and it might even not be commercially viable to grow some crops like coffee. It is estimated that the withdrawal of fungicides will raise the price of food commodities by 13% (Table 5); this would affect the population with lower incomes more. A compromise between a cleaner environment and good cultivation practices has to be found if food prices are to be kept within affordable limits.

Integrated pest management

In an ecosystem that is badly damaged, like that of cotton in India, the first requirement is to reduce the present 20 to 25 rounds of spray to six or seven. In time, reduced sprayings will promote a balance between the pest and its natural enemies. Depending on the crop and the extent of the ecological damage that has occurred, the time to reach equilibrium will vary. It must be appreciated that biological control, although it is an eco-friendly approach, is not a panacea for pest control. Predators and parasites, being biological organisms, are difficult to mass produce, transport, deliver in time, and they struggle in their new environment, compete, co-evolve, disperse, migrate and hence are extremely difficult to use to enforce total control on insect pests and diseases. Success in IPM and biological control will come only when the problem solving is decentralized. Small entrepreneurs with adequate knowledge of plant protection should come forward to start IPM enterprises. Such individuals would need a plant protection retainership of 1000 farm families with a captive operational market to cover another 10 000 ha under IPM/biological control. If, in addition, they undertake ancillary activities such as bee-keeping or mushroom-growing, then a self-sustaining county level business can be established. In the long run, investing in these programmes will benefit the farming community. Detailed project planning on this topic is required to inform the concerned authorities of the untapped rural employment potential that is hidden in IPM.

TABLE 5   Possible effects of withdrawal of fungicides from agriculture on prices

Crop	Expected % increase in the prevailing price
Apples	89
Peanuts	70
Lettuce	41
Carrot	80
Coffee	Would disappear
Banana	Would disappear
Total	13

From Kristen (1990).

The 1990s, I am sure, will see more and more IPM. In our attempt to contain losses caused by diseases, lessons on good agronomic practices will be learned from traditional farming procedures and blended with modern farming systems. The time-tested cropping sequences, companion crops, green manuring, application of farmyard manure, solarization by way of summer ploughing, etc, should be re-investigated to identify good agronomic practices for inclusion in the IPM package.

References

Anjaneyulu A 1986 Virus disease of rice in India. Trop Agric Res Ser 19:14–19

Bandyopadhya AK, Das MK, Ghosh JK, Mondal PK, Pawar AD 1986 Studies on faunistic survey of the natural enemy resources of rice leaf hoppers and plant hoppers of Burdwan district in West Bengal. In: Mukhopadhyay S, Gosh MR (eds) Proc. rice hoppers, hopper borne viruses and their integrated management. Bidan Chandra krishi Vishura Vidyalaya, Kalyani, West Bengal, India, p 122–136

Geddes AHW, Iles M 1991 The relative importance of crop pests in South Asia. Natural Resources Institute (NRI Bull 39)

Duttamajumder SK, Singh N, Agnihotri VP 1990 Behaviour of *Colletotrichum falcatum* under waterlogged condition. Indian Phytopathol 43:227–229

Jayasankar NP, Bavappa KVA 1986 Coconut root (wilt) disease: past studies, present status and future strategy. Indian J Agric Sci 56:309–328

John VT 1968 Identification and characterisation of tungro, a virus disease of rice in India. Plant Dis Rep 52:871–875

Kristen P 1990 Calming suspicious minds. Futures, Michigan State University, Michigan, MI 8:5–13

Nagarajan S 1992 New opportunities and challenges in the business of pest management. In: Survey of Indian agriculture. The Hindu, Madras

Nagarajan S, Joshi LM 1985 Epidemiology in Indian subcontinent. In: Roelfs AP, Bushnell WR (eds) Cereal rusts. Academic Press, New York, vol 2:371–402

Rao MK, Krishnan P 1991 Epidemiology of blast (EPIBLA): a simulator model and forecasting system for tropical rice in India. In: Rice blast modelling and forecasting. International Rice Research Institute, Manila, p 31–38

Raychaudhuri SP, Mishra MD, Gosh A 1967 Preliminary note on transmission of a virus disease resembling tungro of rice in India and other virus-like symptoms. Plant Dis Rep 51:300–301

Sidhu DS, Byerlee D 1991 Technical change and wheat productivity in the Indian Punjab in the post-green revolution period. CIMMYT, Mexico (Work Pap 92-02)

Singh DV, Srivastava KD, Aggarwal R, Nagarajan S, Verma BK 1991 Mulching as a means of karnal bunt management. Dis Res 6:115–116

Singh N 1988 Spread of red rot pathogen and infection from the primary focus in standing sugarcane crop. Indian Phytopathol 41:253–254

Solomon JJ, Govindankutty MP, Neinhaus F 1983 Association of mycoplasma-like organisms with the coconut root (wilt) disease in India. Z Pfl Krankh Pfl Schtz 90:26–32

Vanderplank JE 1963 Plant diseases: epidemics and control. Academic Press, New York

Zadoks JC 1992 Crop losses caused by diseases, insects and weeds in the world: the cost of change in plant protection. In: The costs of change in plant protection. J Plant, Prop Trop 9:151–159

DISCUSSION

Kenmore: You suggested a coverage of one IPM specialist per 10 000 ha. Somehow you have to build in what Moni Escalada called a multiplier effect. If you have 10 people employed producing things, such as biopesticides, you are also going to need outreach staff. In our experience, one IPM specialist per 1000 families is not enough in terms of supporting those families and getting information to them in a timely way. You need probably a 10-fold multiplier effect for your specialists. That may come from several channels: it may come through NGOs, through the extension system at state level, or through commodity marketing groups in a cash cropping area. From our experience, 1:1000 is technical backstopping; 1:100 is actual implemention of advice.

Nagarajan: What I envisage is an enterprise run by a local entrepreneur. Such an individual must have a certain amount of profit, which he is able to reinvest as capital. If one IPM specialist covers 10 000 ha under his business, including consultancy, bee-keeping, sale of mushroom spawn, biocontrol agents, pest diagnosis and IPM aids, he can generate employment and make sufficient profit to run the company. The agricultural universities will have to promote the people with a taste for doing IPM business and give them a brief orientation. Possibly, the lending agencies will provide financial support to start the venture. At the moment only biological control is being taken up as an enterprise; this has to be diversified to include all aspects of IPM.

Kenmore: Sesamia inferens is a problem in sugarcane, which may affect mixed rice/wheat crops. Do you see this? Could there be a source population coming out of sugarcane?

Nagarajan: Sugarcane is a major crop in the Indo-Gangetic plain but *S. inferens* is not a concern. Because it is a major rice pest, one occasionally observes it affecting wheat: the rice/wheat rotation may aggravate this situation.

Jayaraj: Dr Nagarajan mentioned the telescopic cropping system developed in and around Madras, in the southern districts of Andhra Pradesh and northeastern districts of Tamil Nadu. Here there are many pest and disease problems, particularly rice tungro virus, because of the practice of growing rice in three telescopic seasons—two short seasons of 105 days; one medium season of 135 days. Within each season there are again early, mid and late seasons, so there are virtually nine seasons per year.

Another problem is that the irrigation system in this area consists predominantly of tanks. So the vectors are spread not only by the wind, but also through water to contiguous areas of rice fields. All these factors are responsible for repeated outbreaks of tungro and other diseases.

13% of pesticides used in India are used in these four districts. If we solve the problem there, we will be able to prevent further spread down south. Dr Krishna Reddy from the Directorate of Rice Research, Hyderabad, has quantified this. The vector starts in the Godavari delta, comes down to the Krishna delta, then Palar and then goes to the Cauvery delta. So if you want to promote sustainability in rice production, management of rice tungro virus in this area is very important.

The vector species is a complex problem in the coastal areas of Andhra Pradesh and Tamil Nadu. We have four species of green leaf hopper. The greater the population of *N. virescens*, the greater the problem with tungro virus. In the coastal areas and in the adjoining North Arcot district, particularly, nearly 90% of the vector population is *N. virescens*. In the interior districts of the state, *N. nigropictus* predominates; there is little *N. virescens*. If we can identify the vector species properly, and index the degree of tungro virus infection in rice by means of potassium iodide tests and so on, we may be able to promote sustainability in rice production.

Varma: In West Bengal very good work has been done on the epidemiology of rice tungro virus (Mukhopadhyay 1986), demonstrating movement of the virus through the vector in the region. This raises the question of regional management of a problem. Dr Shanker Mukhopadhyay has written a very good report on this.

Royle: A point of information about apple scab forecasting. Horticulture International in the UK has produced an impressive forecast model for apple scab, which is substantially better than the traditional Mills period table. This is being incorporated into scab forecasters.

Jeger: Many forecasting schemes depend on the types of systemic fungicides available and their curative activity. How does the warning scheme that you are operating for apple scab tie in with the types of fungicide that are available and produced in India?

Nagarajan: Most of the fungicide used for apple scab management is either produced or formulated in India. A sequence of protective and systemic fungicides is used to check apple scab. For almost every 50–100 km in the apple-growing state of Himachal Pradesh, there is a checkpost where the leaf litter is examined for maturity of ascospores and the Agromet monitors the weather and feeds the information into the apple scab predictor. Once the period for infection is completed, a siren is sounded. This covers about 20 km and warns the horticulturalists that it is time to spray. Before the season starts, the horticulturalists are informed of the sequence of fungicide to be used. Not all of them follow this, but most of them do. It has worked very effectively and is an ongoing programme in the state.

Royle: I'm encouraged to see the extent to which diagnostic immunotechniques are being used to define pathotypes. I believe there is a great deal of opportunity in some of these methods, not only to characterize pathotypes and strains of different pathogens, but also to identify pathogens in complex situations. More importantly, the future is wide open for using these techniques for measuring disease in new ways. Traditionally, we pathologists measure disease by visual means, often subject to a large degree of error. There are now opportunities to use serological techniques to quantify the amount of pathogen in tissue. This presents real prospects for redefining disease thresholds and even re-evaluating the impact of diseases on crop losses, which will make significant contributions to IPM systems.

I'm interested in the biological control methods used in India. Are these confined to soil-borne diseases, as most are for control of fungal pathogens, or do you have opportunities for biocontrol of airborne diseases, maybe by hyperparasites, in this country?

Nagarajan: We are now starting to look at diseases other than soil-borne diseases. Seed-borne diseases such as loose smut of wheat can now be controlled biologically using specific strains of *Trichoderma viridi*. This produces an antibiotic that the root system is able to absorb; it may then inhibit the mycelia so that the apical primordia don't get infected in time. The percentage of infected heads in biologically controlled plots is lower, and is comparable to that in plots sown with seeds treated with carbandazim.

With regard to leaf diseases, manipulating the leaf surface microflora is very difficult, because generally the weather in north-west India is dry. Leaf moisture is needed to manipulate the leaf surface microflora.

Varma: Some important work has been done on the biological control of bacterial diseases. *Xanthomonas campestris* pv *malvacearum* causes black gram of cotton. Phylloplane bacteria have been found to be very effective in reducing its population. It is a question of having the right humidity and sufficient time for enough bacteria to accumulate. This work has shown very good progress (Verma et al 1983).

Royle: The basis of the western European concepts of integrated farming system scenarios, which were pioneered in The Netherlands and now extend to Germany, the UK and other countries, is the interactions that occur between many of the factors that can be manipulated in agricultural systems. For arable field-type crops, there are many factors which alone do very little, but which in interaction profoundly affect opportunities for control of diseases, pests and weeds. One of the keys in the UK and German systems is the way we till the soil; especially the adoption of 'non-inversion tillage techniques'. Here, the residues of a previous crop, often a cereal crop or a crop with a high degree of biomass, are not taken off or burnt (which is now legislated against in western Europe), but are incorporated into the top layers of the soil, so that about 70% of the residues are incorporated into the top 10 cm with about 30% left on the surface or protruding from it. This gives much more trash than in conventional tillage, which can offend traditional western farmers concepts of clean agriculture. This method has introduced some new considerations for the control of pests. For example, beneficial organisms that attack and prey on aphids accumulate in this situation, though it's not known how. Aphid and thereby viral control is thus encouraged more than with conventional ploughing. However, this benefit is offset by adverse effects—slug populations increase with non-inversion tillage. So there is a counterbalancing situation.

When considered in the context of the entire farming system, there are fascinating opportunities for manipulating the systems towards biological and integrated control. Is this sort of concept feeding through in your ideas for IPM in India, where you could use husbandry techniques? You referred to a few practices, such as modifying sowing dates, but are you looking at them interactively?

Nagarajan: The major concern in Punjab today is the fall in the amount of organic matter in the soil. The lack of a carbon source influences the population of saprophytic fungi, the rhizosphere microflora and ultimately the efficiency of biocontrol. Crop residue management has been found to be useful in sugarcane. For centuries, farmers in Uttar Pradesh have done trash mulching between the sugarcane alleys to check weeds and conserve soil moisture.

Swaminathan: Certain diseases in India have long evaded a proper scientific understanding and solution. Dr Nagarajan referred to coconut wilt; there are also mango malformation, sandal spike and citrus die-back. These are all names which indicate a symptom, and the causal factors are several. Nevertheless, the public perception is that if the scientists really want to, they should be able to find an answer. In Kerala, the Coconut Research Centre at Kayangulam is blamed for not working hard enough, because no solution to coconut wilt has been found. In practical terms, *ad hoc* suggestions were given from time to time. At one stage it was felt that dwarf x tall hybrids might be more tolerant. An interesting feature is that right in the epicentre of coconut root wilt, certain palms give 200–300 nuts per year. If clonal propagation methodology could be perfected, such presumably resistant lines could be propagated. This may be an answer.

Mango malformation is quite a serious problem. In Pakistan, Mr Malik Khuda Baksh Bucha, former Minister who has a beautiful mango orchard near Lahore, told me the same thing. He said: the scientists are always giving me an explanation for the disease, but I want a solution to the problem not an explanation! Maybe, where there are multiple causes for a particular phenotype, which is called a wilt or malformation, an integrated management system would be particularly appropriate.

Nagarajan: Die-back syndrome in citrus and orchard crops is due to complex interactions. Historically, it was attributed to a single cause; that's probably one reason a viable solution never emerged. Proper nutrient management, water management and orchard management are now considered to be a prerequisite to control the citrus decline, which is caused primarily by *Phytophthora* spp. An interdisciplinary research programme is needed to develop strategies to contain perennial crop diseases.

Varma: There are very useful recommendations for integrated management of mango malformation (Varma 1983). It is a management problem; the disease is caused by a fungus which is spread by a mite.

Sandal spike has been well demonstrated to be caused by a mycoplasma-like organism. Like any other disease caused by such organisms in perennial plants, it's again a question of management. Good practices have been recommended (Raychaudhuri & Varma 1980). For forest tree management it is rather difficult, but it is a manageable problem.

I have not myself worked on coconut root wilt. It has been shown that it is possible to improve the productivity of trees by proper nutrient management. Mycoplasma-like organisms have also been found to be associated with this disease.

Jeger: Many diseases with a complex aetiology show the phenomenon of 'convergent symptomatology' in different parts of the world. Mango malformation may be one example. Although the aetiology is well worked out in the Indian subcontinent, in other parts of the world there are similar problems which may be due to other causes. The same applies to some of the problems in coconut and cloves.

References

Mukhopadhyay S 1986 Virus diseases of rice in India. In: Varma A, Verma JP (eds) Vistas in plant pathology. Malhotra Publishing House, New Delhi, p 111–122
Raychaudhuri SP, Varma A 1990 Sandal spike. Rev Plant Pathol 59:99–107
Varma A 1983 Mango malformation. In: Singh KG (ed) Exotic plant quarantine pests and procedures for introduction of plant materials. ASEAN PLANTI, Malaysia, p 173–188
Verma JP, Singh RP, Chowdhury HD, Sinha PP 1983 Usefulness of phylloplane bacteria in the control of bacterial blight of cotton. Indian Phytopathol 36:574–577

Sustainable agriculture and integrated pest management in China

Shimai Zeng*

Department of Plant Protection, Beijing Agricultural University, Beijing 100094, China

Abstract. In developed countries, emphasis is being switched from high productivity through the use of high inputs to ecologically sustainable agriculture. In developing countries such as China priority must be given to increasing food production while simultaneously trying to optimize sustainability. Achievements in plant protection are being countered by continued evolution of the pest ecosystem, in part driven by application of pesticides or the introduction of new crop varieties. Future management of the agricultural ecosystem requires the development of a method of 'super-long-term' prediction to evaluate possible consequences of different strategies of plant protection. Crop plants with durable resistance to pests must be derived by conventional breeding or by using biotechnology and genetic engineering. Genetic vulnerability can also be reduced by techniques such as gene rotation and mixed cropping. Biological control of plant pests shows promise but requires ecological study of the relationships among crop, pest and natural enemy. Implementation of sustainable pest management will need training and education of farmers, extension workers and policy makers to deliver new information in the developing countries.

1993 Crop protection and sustainable agriculture. Wiley, Chichester (Ciba Foundation Symposium 177) p 228–232

The general perspective of sustainable agriculture is common to all countries, but the strategic points and suitable approaches might differ according to local conditions. In some developed countries, especially those that have suffered from surplus agricultural production and environmental damage caused by high energy/input agriculture, ecological protection is emphasized and low input-low output sustainable approaches, such as 'ecological agriculture' or 'organic agriculture', are suggested. In China, a developing country, priority must be put on increasing food production, and an approach that combines high output, high efficiency and sustainability must be pursued. That is, high input and extensive cultivation should be ingeniously designed to ensure not only high yield per unit area of cultivated land and high efficiency, but also to reduce

---

*Unfortunately, owing to illness, Professor Zeng was unable to attend the symposium.

detrimental environmental side-effects to as low a level as possible. In general, high output inevitably requires a corresponding high input, but improving the structure of the inputs might be more favourable than just increasing the quantity. Inputs for agriculture consist of materials and energy, i.e. fertilizers, water, pesticides and agrochemicals, mechanization, solar radiation and heat, as well as scientific and technological input, i.e. cultivars, cultural measures and management options. In sustainable agriculture, scientific and technological input must be emphasized, to maintain a high growth rate of production while reducing the harmful side-effects of high-input systems. Accordingly, integrated pest management in the future should be more ecologically orientated and should depend more on multiple and durable host resistance, well adjusted cultural practices and ecologically sustainable biocontrol measures.

Human-guided evolution of the pest ecosystem

In China, great achievements in plant protection are accompanied by a continuing struggle against pest problems. Besides the resurgence of some old pests, the introduction of some foreign pests and the emergence of new ones, the interaction or action–reaction between humans and the pest ecosystem seems to be everlasting. Two examples are given below.

The ascendance of pests that were formerly of little importance

In North China, from 1950–1991 the most important disease of wheat in successive periods has been stripe rust, leaf rust and powdery mildew; on corn, it has been southern and northern leaf blight, head smut, dwarf mosaic virus disease, ear rot and foot rot, and *Rhizoctonia* sheath blight. The principle factor inducing a change in the pest pattern is the spectrum of pest species against which a cultivar is resistant, coupled with the effects of high density planting and increased application of nitrogenous fertilizer and irrigation. Kernel smut and false smut of rice, formerly of little importance, have become more and more severe in some places after the wide release of hybrid rice that is more vulnerable to these diseases. They infect the flowers, and the new hybrids have a longer flowering time of the male sterile parent. Rossette stunt of wheat transmitted by *Laodelphax straitellus*, a viral disease that has never been severe in winter wheat regions under a monocropping system, became a serious problem where a 'wheat–corn–millet or sorghum' intercropping regime (three crops per year) was popularized. Similarly, a wheat–corn double-cropping system has increased the threat of wheat scab and has caused several outbreaks.

Boom-bust cycle of resistant cultivar and of pesticide

Between 1950 and 1992, five groups of important wheat cultivar lost their resistance to stripe rust, because of the evolution of new virulent races. The blast resistance of rice cultivars has lasted no more than 3–5 years in many mountain regions in South China and coastal regions in North China. Resistance to many modern insecticides has developed in many kinds of important insects. Owing to rapid development of resistant strains of the fungus, Rhidomil is no longer a highly effective fungicide to many important Oomycetes in China.

From the above examples, it is clear that in most instances, pest disasters are induced by human activities and are the ill effects of imperfect agricultural activities in farming system reformation, cultivar replacement, cultural practice improvement, or the adoption of new pesticides. In general, the variation of climatic conditions from year to year affects only the fluctuation of the population dynamics of pest species. Agricultural activities can often change the community structure of the pest ecosystem, thereby influencing its evolution. Man-made pest problems have occurred frequently and we learned to solve them always only after the event, just as the old saying in China: 'Let him who tied the bell on the tiger take it off'. Better, we try to achieve 'a fall into the pit, a gain in your wit'.

Super-long-term prediction of the evolution of pest ecosystems and large-scale ecosystem management

In the future, artificial evolution of the agro-ecosystem might proceed more rapidly than ever, and along with it, various new pest problems would inevitably occur. Therefore, it is necessary to carry out a study on 'super-long-term prediction' of the pest ecosystem to try to forsee the evolutionary trends of the pest ecosystem, the risk of pest problems possibly induced by the extension of some new cultivar or agricultural technique, and the possible consequences of different strategies of plant protection. Macroscale pest management needs to prevent potential disasters, at least some man-made disasters, as early as possible. So far, the theoretical basis of pest prediction, both short and long term, is the population dynamics of pests under comparatively small scales of time and space, whereas that of the super-long-term prediction is the process and mechanism of the evolution of the pest ecosystem. While there may be indirect influences of socio-economic factors and the strong impact of global climate changes, such as the greenhouse effect, the major determinants of the dynamics of the whole pest ecosystem are agricultural activities. These activities, being the controllable input of the system, could be adjusted or regulated to a greater or lesser extent by people in systems management. So, a macroscale, long-term system of management could progress step by step as the consequence of a rolling super-long-term prediction. This study will provide the basic framework of a strategy for IPM in sustainable agriculture. System analysis will be an important tool in this study.

Durable resistance to pests and perpetuating host population resistance

For the adoption of sustainable agriculture, host plant resistance to pests must be strengthened. For diseases caused by mutations in single genes (and for insects to which resistance is encoded by a single gene), e.g. wheat rusts and rice blast, high resistance is not very difficult to breed, but it is often short lived. For other diseases (insects), e.g. wheat scab and rice sheath blight, the observed partial resistance seems to be or is mostly expected to be durable but is often not strong enough to satisfy the needs of plant protection: it is very difficult or might even be impossible to breed highly resistant cultivars by means of conventional breeding methods. Biotechnology and genetic engineering will be powerful tools, but still there remains the key problem: how to assure that the introduced resistance will be durable. So far, experience has shown the utmost importance of basic research on the molecular mechanisms of resistance and host/pathogen interaction to bring out the full role of genetic engineering. Research on durable resistance must be broadened and deepened, from the concept and definition, to the biochemical and molecular basis, from method of identification to method of breeding.

Genetic vulnerability to pests arises from the genetic homogeneity of crops in modern agriculture. Theoretically, population resistance of host to pest can be stabilized by means of host genetic re-diversification. Many strategies have been suggested, such as regional gene deployment, gene rotation, gene pyramiding, use of multiline, multi-genotypic varieties or mixed varieties, and mixed cropping. None of these has been practised satisfactorily in China owing to many technical difficulties, economic limitation and/or lack of suitable knowledge. Genetic re-diversification of crops involves scientists, farmers, extension workers, merchants and consumers. The study of effective strategies involves agronomy, plant breeding, plant pathology, entomology, agriculture management and even government policy. High-level, multidisciplinary study must be placed on the agenda.

Biocontrol in a narrow and a broad sense

Biocontrol must play a more active role in sustainable agriculture. The effectiveness and durability of it are determined by many factors—ecological, biological, technical and economic ones. On the basis of limited experiences, it seems that combinations of pests and natural enemies, including parasites, that do not have a long history of co-evolution can be used successfully, but more rigorous ecological studies are urgently needed. No less important is the protection of natural enemies and parasites, either native or exotic. In China, dramatic increases in the populations of many insect pests have been induced just by accidental killing of the relevant natural enemies and parasites by wide-spectrum pesticides, such as rice brown plant hopper and leaf folder by DDT and 666. Hence, any measures, such as rational or need-based application of pesticides, the choosing of selective pesticides and the lowering of the action threshold, which favour the protection of natural enemies and parasites should be studied further and viewed seriously as a necessary component of biocontrol in the broad sense. The application of antibiotics is considered in the scope of biocontrol in its broadest sense by some people, but in reality antibiotic preparations via fermentation or artificial synthesis are similar to general chemical pesticides and might possess the same ill side-effects, thus being no longer an ecologically sound means of pest control. The possible development of resistance of target pests to living biocontrol agents needs special attention. Once again, ecological study focusing on the three-way relationship among crop, pest and natural enemy influenced by the physical environment forms the theoretical basis of biocontrol. Such study should be pursued vigorously.

Education in pest management

In progressing toward sustainable agriculture, IPM would naturally evolve to sustainable pest management. This is more difficult to deliver by normal extension mechanisms owing to its 'soft' and dynamic nature. Its implementation and extension are much more difficult than those of cultivar, fertilizer, or such singular techniques as pesticide or agrochemical application. It is a knowledge-based extension rather than a material-based one. The lack of awareness about IPM and sustainable pest management of most farmers, many agrochemical businessmen and even some of the extension workers and policy-makers is the socio-cultural constraint limiting the successful development of sustainable pest management. Training and education to deliver the new information as it is developed are of crucial importance in developing countries.

Towards the rational management of the insect pests of tropical legume crops in Asia: review and remedy

John A. Wightman

International Crops Research Institute for the Semi-Arid Tropics, Patancheru, Andhra Pradesh 502 324, India

Abstract. The productivity of legume crops, especially the pulses, has not increased markedly in 30 years. This is a serious matter in this time of exponential human population growth because legume crops provide essential diet components that are not present in cereals at a price that is affordable to the majority. The preferential allocation of research resources to cereal crops is one reason for this stagnation. I suggest that an international effort should be mounted to redress this situation. Most Asian legumes have in common a cadre of insect pests that are reducing yields to levels at which harvesting is not economic. The position is worsening because insecticide resistance in key production areas has rendered many species uncontrollable. The short-term remedy is to persuade farmers to reduce their dependence on insecticides. Long-term solutions should be aimed at a common policy centred on the principal components of integrated pest management (IPM) schemes. Governments should be supplied with information that will permit them to set priorities according to absolute crop loss indicators. These will in turn provide a rational basis for the subsequent development of regional strategies. The establishment of an IPM centre in Asia would support the facilitation of this ideal. Such an organization would become the focal point for collecting, developing and disseminating the information and technology needed to make a quantum leap in legume production.

1993 Crop protection and sustainable agriculture. Wiley, Chichester (Ciba Foundation Symposium 177) p 233–256

There is in Asia a considerable amount of motivation towards developing strategies for integrated pest management (IPM) within the context of sustainable agriculture (see this volume: Jayaraj & Rabindra 1993, Varma 1993, Nagarajan 1993, Escalada & Heong 1993). There is no doubt that this activity will continue to contribute to the story of increasing production, of rice in particular. As such, this will be cited as the major research-led agricultural success story of this century. So successful has it been that several countries in Asia that once imported their staple are now major rice exporters. Not only

233

that, but scientists predict that by 2050 rice productivity will have reached 20 t ha^{-1} per annum from two growing seasons—enough to feed 150 people for one year (*The Australian* 30 April 1992). This is fortunate because data released before the June 1992 Earth Summit predicted that the world's population will double by 2050 (*Guardian Weekly* 3 May 1992). An extrapolation of current world population data (Food and Agriculture Organization of the United Nations) gives one the impression that there will be considerably more people who will want to eat rice and other cereal products by this time:

Population of:	World	50×10^8
	Asia	30×10^8
	China (People's Republic)	12×10^8
	India	9×10^8
	Indonesia	2×10^8

The problem

People, especially children, who eat only cereal products are not likely to be healthy. Such a diet does not include all the amino acids, lipids, vitamins and mineral salts that are necessary for body maintenance and disease resistance. Thus it is customary among those who can grow or afford to buy them to supplement the staple with meat, eggs, cooking oil, fruit and vegetables. The poor and the voluntary vegetarians, who together form the majority of the people of Asia, rely on several of the 20 or so legume crops to supplement their diet. Unfortunately, pulse production in Asia has not kept pace with cereal production or the growth of the human population (Table 1).

In considering future developments of IPM in the context of sustainable agriculture, I have restricted myself to considering legume crops and the farming systems in which they grow because: (1) their under-production in Asia does not bode well for the future well-being of the continent; (2) I believe the production of cereal products is assured into the foreseeable future; (3) I cannot imagine that the current deteriorating situation can be allowed to continue without concerted international action and I wish to encourage this process.

Insect pests are a major contributor to the stagnation in legume production. Not only are most of these crops inherently susceptible to a large and diverse cadre of insects, but insecticide resistance is increasing daily within this cadre. This means that insect pests are becoming more and more out of control as a result of the continuing elimination of natural control processes as more and more insecticide is applied in attempts to control them. Further details and examples are available (Wightman 1989, Legumes Program, ICRISAT 1991): space limitations preclude the discussion of case histories.

The potential tools needed for managing the insects are often well known, but only to scientists and then only on a theoretical basis. What is needed

TABLE 1 Population and production data for Asia

		1961	1966	1971	1976	1981	1986	1991
Population ($\times 10^{-9}$)		1.9	2.0	2.1	2.4	2.6	2.9	3.0
Cereal	Area (Mha)	239	249	260	268	272	271	306
	Production (Mt)	248	309	385	445	528	626	823
	Productivity (t ha^{-1})	1.04	1.24	1.48	1.66	1.94	2.31	2.69
	Per capita (kg)	131	154	183	185	203	215	274
Pulses	Area (Mha)	37	34	32	33	32	33	36
	Production (Mt)	23	19	29	21	20	22	25
	Productivity (t ha^{-1})	0.62	0.56	0.62	0.64	0.62	0.67	0.69
	Per capita (kg)	12.1	9.5	13.8	8.8	7.8	7.6	8.4
Pulse: cereal production (%)		9.3	6.2	7.5	4.7	3.8	3.5	3.1
Groundnut	Area (Mha)	9.2	10.6	10.8	10.4	11.4	13.0	13.0
	Production (Mt)	7.5	8.4	10.4	9.1	13.4	15.3	15.4
	Productivity (t ha^{-1})	0.82	0.89	0.96	0.87	1.18	1.18	1.18
	Per capita (kg)	3.9	4.2	4.0	3.8	5.1	5.3	5.1
Soybean	Area (Mha)	11.3	9.8	9.2	8.2	10.0	11.7	13.5
	Production (Mt)	7.0	9.2	9.8	8.1	11.2	14.7	16.0
	Productivity (t ha^{-1})	0.62	0.94	1.07	0.99	1.12	1.35	1.19
	Per capita (kg)	3.7	4.6	4.7	3.4	4.3	5.1	5.3

therefore is emphasis on on-farm implementation studies, technology evaluation and integration. In this case 'integration' includes breaking down the walls between scientific disciplines.

Tropical legume crops

Most of Asia lies in the tropics, but the legumes associated with the cooler regions —lentils, faba beans and sweet (green) peas—are included in the list of food legumes (Table 2). A description of the distribution and uses of legumes requires a treatise by itself (Purseglove 1968, Smartt 1976, ICRISAT 1989). In general, in East and South-East Asia legumes are most likely to be eaten as green vegetables. Green gram and soybean are often processed in the manufacture of noodles and fermentation products. In predominantly vegetarian South Asia, nearly all the legumes listed are eaten, but the pulses (dried seeds) are significant as supplements to the staple rice and wheat dishes. Groundnut and soybean are dominant sources of high quality vegetable oil but are also eaten in many forms.

Production levels have not approached the 1:9 pulse:cereal ratio (11% pulse) considered normal for Asia for many years (Table 1). The desirable 3:7 ratio (43% pulse) may never have been achieved on this continent (Hulse 1990). The production of legume oil seeds (groundnut and soybean) presents only a slightly brighter picture in terms of increasing productivity and the area sown. All production figures are not gloomy: soybean and groundnut production are increasing in India (FAO data) and Indonesia (Broto & Bottema 1990) (Table 1).

Current yields for most of the pulse crops are in the region of $0.4–0.6\,t\,ha^{-1}$. Fortunately, the achievable yield may be more than five times that amount, or more for chickpea, soybean, mung and groundnut (Smartt 1976). This huge yield gap indicates that there is at least the potential for considerable management-led impact on legume production in the region.

The conclusion we can draw so far is that the food-energy demand of the more disadvantaged sectors of Asian communities may be fulfilled into the next century. Despite this, malnutrition will be among the foremost factors that will sap the ability of the huge cadre of Asian labourers ('the poor') to grow staple supplements or earn the money to buy them. The inability of the poor to buy cereal supplements was already apparent 12 years ago (Bidinger & Bhavani 1980). The situation is getting worse as the Asian industrial revolution draws farming people from the land to the cities. Stable supply in the face of increasing demand is already leading to price hikes. The price of 1 kg of split pigeonpea (dahl) in southern India has increased from less than half a labourer's daily wage to more than a full day's pay in three years.

There are many reasons food legume production in Asia has stagnated for such a long time, for instance:

(1) The comparative lack of national and international investment in quality research that is directed at increasing legume productivity. There is certainly

an imbalance in resource allocation that is probably associated with the politically sensitive nature of crops such as rice.

(2) The release of new varieties bred for high yield potential on research stations with high susceptibility to several pests in low-input conditions. Also the converse—a lack of varieties adapted to the biotic and abiotic features of specific environments and the needs of farmers.

(3) Slowness in the approval, release, multiplication and distribution of improved varieties by national programmes.

(4) Prestige—'real farmers grow rice'.

(5) Considering legumes as adjuncts to cereal-based systems, not as primary components.

(6) High cost, e.g. groundnut, and poor storage prospects of seed (mainly because of bruchid beetles—pests of all legume crops).

(7) The inherent intractability of the production problems (including insects) that result from the complexity and fragmentation of the farming systems, the often harsh environment and the number of legume species involved.

The insect pests

The core of the production problems lies in the last constraint and the associated implications for the farm economy. The most important production problem (after uncontrollable abiotic factors like drought and flooding) is a cohort of insect pests (Table 3). Its members regularly reduce the yield of legume crops to levels at which it is not in the interest of the farmers to harvest them. However, other components of the farming systems in which these crops grow are also involved. For instance, in India and Thailand, the heavy application of insecticides to cotton has enhanced the injury caused by *Helicoverpa* spp. and whiteflies to legume crops in neighbouring fields.

The problem created by these pests is different and considerably more intense than it was perhaps 10 years ago. Many insects are now out of control despite (because of) the liberal application of insecticides. The best documented example within the current context is that of *Helicoverpa* resistance to several insecticide classes in Andhra Pradesh (Table 4). A resistance factor of 10 is sufficiently high to consider implementing resistance management schemes. The data show that this has been exceeded greatly in recent years.

Specialists from the Philippines, Vietnam, Indonesia, Thailand, Sri Lanka and India have reported that farmers apply insecticides to legume crops at least once a week to kill *Maruca*, *Etiella*, *Helicoverpa* and the large Heteroptera. The natural control process has been destroyed and the target insects are highly resistant to pesticides. Marketed produce must be highly contaminated with pesticides. Farmers can see no answer to the problem except to apply more insecticides—the insecticide treadmill (Legumes Program, ICRISAT 1991).

TABLE 2 Food legumes of Asia

Species and common names	Where commonly grown	Comments
Arachis hypogaea peanut, groundnut	Throughout Asia	Oil, food and process crop
Cajanus cajan pigeonpea, red gram, tur	India	Staple pulse
Canavalia ensiformis jack bean	Dispersed thinly throughout Asia	Low density pulse and green manure
Canavalia gladiata sword bean	India	Climbing habit; green pods and immature seed consumed fresh
Cicer arietinum chickpea	North Asia, northern India, Myanmar	Staple pulse; cold adapted, drought adapted
Cyanopsis tetragonoloba cluster bean	Southern Asia	Vegetable, fodder and green manure
Dolichos uniflorus horse gram	Southern India	A poor person's pulse and stock food
Glycine max soybean	Throughout Asia	Important oil and process crop, also grown as a fresh vegetable
Lathyrus sativus grass pea, chickling pea	Mainly India	A poor person's crop; some genotypes have dangerous levels of a cumulative toxin
Lens esculenta lentil	North Asia, temperate China and at high altitudes in the tropics	
Phaseolus aconitifolius moth bean	Mainly India	Adapted to hot, dry conditions
Phaseolus angularis adzuki bean	East Asia	Grown in rice stubble; short-day adapted
Phaseolus lunatus lima bean, butter bean, Burma bean	Myanmar	Bush and climbing forms; widely adapted
Phaseolus vulgaris kidney bean, common bean, french bean, snap bean, string bean (etc)	Throughout Asia	Green vegetable and pulse

Species	Distribution	Notes
Pisum sativum green pea, sweet pea	Throughout Asia	Temperate climate required (>1200 m in the tropics)
Psophocarpus tetratonolobus winged bean, Goa bean	South-East Asia	Multipurpose crop adapted to moist tropics
Vicia faba Horse bean, faba bean, tick bean, broad bean	Northern Asia and temperate China	Vegetable and pulse crop
Vigna aureus green gram, mung	Throughout tropical Asia	Of great importance, eaten in many forms
Vigna lablab hyacinth bean, lablab bean	Mainly India	Climbing or bush forms of value as pulse and fodder; grows in a range of environments
Vigna mungo black gram	Mainly India, but also parts of South-East Asia	Eaten as a pulse or used as flour
Vigna sesquipedalis yard long bean	South-East and East Asia	Favoured vegetable crop
Vigna umbellata rice bean	Throughout Asia	Follows rice or sown to climb over mature maize plants
Vigna unguiculata cowpea	Throughout Asia	Not as popular as some other pulse crops, perhaps because of sensitivity to stem flies
Vigna subterranea Bambara groundnut	Only in South-East Asia	Of great potential because of its resistance to pests and tolerance to drought, heat and poor soils

Sources: Purseglove (1968), Smartt (1976).

TABLE 3 The most serious insect pests of legume crops in Asia

The insects	Distribution	Crops	Comments
HOMOPTERA			
Aphis craccivora groundnut or cowpea aphid	Cosmopolitan	Most, if not all	Spread viral diseases, reduce plant turgor and remove essential nutrients from the plant
Aphids in general	Cosmopolitan	All	Many species of aphid carry non-persistent viruses that are transmitted by 'casual' probes of non-normal hosts
Bemisia (*tabaci*) whiteflies	Cosmopolitan	Most	Spread viral diseases, reduce plant turgor and remove essential nutrients from the plant
Empoasca spp. and other leaf hoppers (jassids)	Cosmopolitan	All	Spread viral diseases, reduce plant turgor and remove essential nutrients from the plant
HETEROPTERA			
Large bugs (Coreidae, Pentatomidae, Lygaeidae)	Cosmopolitan	All	By feeding directly on the developing pods these insects reduce seed yield at low densities. Many are large (>3 cm long) in the adult stage and difficult to kill with insecticides
especially: *Anoplocnemis phasiana* *Clavigralla* spp. *Nezara viridula* *Piezodorus* *Riptortus* spp.	India	All	
Small bugs (Miridae) *Creontiades* sp. *Eurystylis* sp. *Campylomma* spp.			Destroy flower buds, have economic impact at low densities (e.g. one per groundnut plant)
THYSANOPTERA (thrips)			
Indeterminate species	Indonesia	Mung	Can eliminate seed yield of unprotected plants if crop sown at wrong time
Thrips palmi	India	Groundnut	Vector of bud necrosis virus

COLEOPTERA

	Distribution	Crops	Notes
Bruchidae, including			A group of beetles associated with legumes before and after harvest: the most important storage pest of legume seeds
Callosobruchus spp.	Cosmopolitan	All	
Caryedon serratus	South Asia	Groundnut	
Acanthoscelides obtectus	Wide	Many	
Scarabaeidae white grubs	Cosmopolitan	Groundnut, chickpea	Major root pest of groundnut (and chickpea in Myanmar) but may attack other crops

DIPTERA

Agromyzidae	Distribution	Crops	Notes
Ophiomyia phaseoli bean fly	Cosmopolitan	*Vigna* spp. *Phaseolus* spp.	The major pest of seedlings
Melanagromyza obtusa pod fly	India, Thailand Vietnam	Pigeonpea	Larvae live in pod and destroy seeds

LEPIDOPTERA

	Distribution	Crops	Notes
Helicoverpa spp. especially *armigera*	Cosmopolitan	Probably all	Major pest throughout the continent; pod borer and seed eater
Maruca testulalis	Cosmopolitan	Most, not soybean	Lives in more humid areas; destroys flowers and young pods
Etiella zinckenella	Cosmopolitan	Probably all	Eats developing seeds
Lampides boeticus	Widespread	Many	Eats flowers and green pods
Amsacta spp. and other 'hairy caterpillars'	Southern Asia	Many	Polyphagous defoliators
Aproaerema modicella	Widespread	Soybean, groundnut	Leaf miners; major yield reducers but sporadic
Spodoptera litura	Widespread	Many crops	Polyphagous; mainly a defoliator but can attack developing pods

Sources: Singh (1990), Marwoto & Neering (1989).

TABLE 4 Levels of resistance to insecticides in *Helicoverpa armigera* taken from pigeonpea, chickpea and cotton in Andhra Pradesh and Tamil Nadu, South-East India

	Number of samples	Resistance factors[1]	
		1	2
1989–90			
Pyrethroid (cypermethrin)	7	27 (1–100)	572 (20–2100)
Organophosphate (quinalphos)	6	—	2 (2–4)
Carbamate (methomyl)	1	—	8
1990–1991			
Pyrethroid (cypermethrin)	23	12 (0.5–64)	157 (7–830)
Organophosphate (quinalphos)	4	—	3 (2–4)
Carbamate (methomyl)	3	—	15 (6–30)
Chlorinated hydrocarbon (Endosulfan)	3	4 (2–7)	—

Summarized from data of Armes et al (1992).
[1]Materials were applied topically to 30–50 mg larvae. 1, control strain from New Delhi; 2, control strain from Sudan. Data are the mean (and range) of samples that usually consisted of more than 200 larvae.

If legumes are to remain part of the diet of Asians something has to be done about these insects and the insecticides—now.

Rational pest management

I shall add little new to what has already been written by many erudite people (e.g. Brader 1979, Kenmore et al 1987, van Emden 1980) about rational approaches to pest management and related research in developing countries. I hope another rearrangement of some of these realities will not be amiss.

The essence of the problem is to make the investment in legume production a more attractive proposition to farmers—a multifaced proposition. For instance, there is a case for freeing irrigated land for legume production and switching emphasis to the comparatively neglected production of upland (dry-land) rice (Maurya et al 1988). This scheme has merits: it would boost legume production and would make more efficient use of irrigation water (Wightman 1990).

Irrespective of the nature of any new schemes, an immediate reduction in the amount of insecticide applied to Asian legumes and associated crops is needed. This implies the provision of alternative pest management practices that will stabilize the situation at tenable levels of production. In other words, we need to apply the principles of IPM within the context of sustainable agricultural production. As the meaning of 'IPM' and 'sustainable' vary according to the individual and the context, the following definitions pertain to this paper:

A **sustainable** farming system is managed so that:
(1) its long-term productivity and quality and that of its environment are maintained at the status quo or are improved with time;
(2) annual productivity and/or profit are optimized;
(3) between-season variation in productivity and/or profit is minimized at an optimum level.

IPM:

Management activities that are carried out by farmers that result in potential pest populations being maintained below densities at which they become pests, without endangering the productivity and profitability of the farming system as a whole, the health of the farm family and its livestock, and the quality of the adjacent and downstream environments.

IPM activities can be divided into:
(1) growing varieties that have resistance to the pests characteristic of the agroecological zone of the farming system; (2) adopting farm management practices that counter the proliferation of pests but which may foster natural control processes; and (3) applying insecticides when needed in such a way that the objectives of IPM are not impugned.

Farmers, researchers and extension officers also need to develop an understanding of the applied ecology of the organisms involved to be able to evaluate the implications of the management options. It will be noted that these definitions and clarifying comments do not refer specifically to insects, but to pests, a term that covers all biotic crop production constraints. We are largely concerned with insects in this paper, as they tend to be the major biotic constraint in the current context, either directly or as vectors of viral diseases. It is sincerely hoped that 'integration' will involve all disciplines, sooner rather than later.

Insecticide resistance management (IRM) is pivotal to the whole process. There are two aspects that need to concern us: (1) farmers perceive the need to apply more and more insecticide because they cannot kill insects with lower, less frequent applications. This leads to the continuing destruction of natural enemies and the pollution of the target and downstream environments. (2) Cross resistance within and across chemical classes will lead to the withdrawal by the agrochemical industry of whole groups of insecticides, including the so-called 'soft' insecticides that have an important place within IPM schemes.

Immediate solutions: fire-fighting

Without question, farmers can be immediately responsive to demonstrations showing that they can withdraw from the insecticide treadmill without losing their livelihood. Procedures that describe to farmers exactly when they need to apply insecticides to several crops already exist. If followed, they will result in the reduction of insecticide application (e.g. Marwoto & Neering 1989, Wightman & Ranga Rao 1993, Wightman et al 1993).

However, many farmers may be understandably nervous of taking such a step. They will require a substitute for insecticides, perhaps in the form of a resistant crop variety. If so, sufficient seed should be immediately available. Such a variety may not give the highest possible yield but is more likely to give sustained production over time than a high-yielding variety supported by insecticides. The introduction of crop insurance schemes or other government-led incentives for non-sprayers may also be considered.

The simplest management tool is the decision not to grow a susceptible crop for one or more seasons or crop cycles. Group action within a community to switch production to another crop may lead to the decline in the local population of the key pests. The substitute crop could be the insect- and disease-resistant bambara groundnut (that is, admittedly, largely untried and 'undeveloped' in Asia), a green manure crop or a non-legume non-host such as chilli, castor or sweet potato. The implications are that a farming community would have to act cohesively and would have to be prepared to change crops again as problems build up. There would undoubtedly be downstream implications for markets and other infrastructural matters. Should problems arise, governmental support should be at hand to smooth out problems, in the name of sustainability and guaranteed food production within the region.

Longer-term solutions

This covers activities that should be implemented now at a national programme and international level. A suitable time horizon is 5–10 years. The appearance of the following items in this list implies that some of these critical activities are not always the accepted procedural practice.

Government acceptance and support

The first step into the future of legume crops is for governments and their advisors to decide that action needs to be taken to reduce insecticide application. For instance, the Government of The Netherlands has decided to implement a 'Multi-Year Crop Protection Plan' that will, by 2000, halve the amount of insecticides applied to Dutch agricultural land, with a baseline of the 1984–1989 average (Zadoks 1991).

Definition of the legume pest problem in absolute terms

The second step is to define the current problems in terms of the distribution of the insects, their density ranges, the amounts of insecticides applied (and the modality of application) and other relevant practices. Estimates of the reduced yield attributable to these and other constraints are needed to put them into perspective. This will require the initiation of farm surveys in some countries. The collation of the data collected within a central Geographical Information System is a prerequisite for the development of an understanding of the continental pattern of the problem within agroeconomic zones. *It would be a considerable insult to common sense if this process was limited to insect pests.*

This information is needed to guide the planners of the region through the advantages and disadvantages of sowing various legume crops in different seasons or within specific agroecological zones. For instance, chickpea can be grown in many environments without pesticide application. Pigeonpea, especially varieties with less than medium duration, are highly susceptible to insect pests in Thailand, Vietnam, Myanmar, Indonesia, Sri Lanka and India (e.g. Wallis et al 1988, Hong et al 1992). However, pod fly-resistant lines are currently available for relevant environments.

Monitoring the sociological and economic implications of changing farm management patterns to accommodate IPM

In parallel with the above, selected farmers and farms in stricken tracts of land (and those selected for initial implementation studies) should be assessed by economists from the outset to monitor the acceptability and relevance of proposed modifications in farm management. An evaluation of the concepts held by farmers about pests and how they deal with them should be an integral part of this process.

The development of national or regional IPM-related strategies

The third step is for agricultural planners to adopt a general strategy set that is simple and that can be applied, in principle, across the continent. Such a strategy might be based on the following concepts:

The rationalization of insecticide application. The use of insecticides should be managed and hopefully reduced by implementing insect resistance management (which is directly linked to the essence of IPM) (Forrester 1990), whether or not insecticide resistance has been detected[1]. Monitoring high risk populations for insecticide resistance would be a key feature of such a strategy[2]. Steps would thus be taken to ensure that farmers have the best possible advice about when to apply insecticides.

---

[1]As far as I know, this is not happening in Asia.
[2]This is being done at or out of ICRISAT (International Crops Research Institute for the Semi-Arid Tropics) Center, within the limits of staff and budget.

Improvements in the technology and practice of pesticide application should be promoted throughout Asia. One approach to fostering this process is the encouragement of village pest management 'clubs'. In these, farm and pest management strategies are discussed and monitored within the community, with guidance from trained advisors if needed. This mechanism promotes ownership in the technology and synchrony among farm activities.

The exploitation of host plant resistance.   Information concerning host plant resistance to pests should be collated and disseminated. It is necessary to be clear about cases where no or little resistance exists in a crop species and/or the relevant 'wild' species so that priority can be put on the development of other strategies. When resistance to key pests cannot be integrated into the germplasm of a crop, biotechnological techniques could be considered with a view to introducing relevant resistance genes from another species.

Where necessary, breeders should be persuaded that, in order for 'their' products to be relevant components of sustainable agriculture, i.e. in farmers' fields, they should work with farmers and specialists in other areas of agriculture. This process will ensure the production of varieties that farmers recognize as being as or more acceptable than existing varieties, in terms of taste, quality (ease of harvesting, seed colour, oil content), yield and the other uses to which farmers put the crop (e.g. fencing, thatching, fodder). These varieties must be able to survive as a crop in farmers' fields without excessive applications of pesticide (Byth 1980, van Emden 1980). Public or private sector agencies must also be empowered to evaluate, multiply and distribute new varieties effectively.

If these concepts can be applied successfully in practice, natural control processes would have a chance to recover. This, in turn, would mean that other management-related options had more impact.

Promotion of natural control processes.   The degree of biodiversity within farming systems should be maintained or increased by activities such as agro-forestry (provides nesting sites for predatory birds) and selective multicropping and weeding to provide refuges for arthropod parasites and predators. Organisms that cause disease in insect pests should be made widely available through mass production and distribution. Other management techniques that reduce the impact of pests and improve crop production should be encouraged (by direct subsidies). These include the adoption of suitable irrigation procedures and mulching with shiny plastic.

Our ability to make optimum use of biological means of crop protection depends on an understanding of the life cycles of pests and their interactions with crop plants. This can be achieved by: (1) setting up an international network of observers and traps (the nature of which will depend on the species) to document the times of year when attack by pests is most and least likely; (2) determining in more detail the relationships between pest density and crop damage.

It is also important that scientists and extension specialists are able to transfer the information to where it is needed. This requires the establishment of an international communication network of plant protection specialists. There is also a need to develop software that will give real-time guidance to farmers about when to apply pesticides (Wightman & Ranga Rao 1993) and to IPM researchers about the best strategies to adopt given specific levels of host plant resistance, natural control and pesticide efficacy for a range of pests and crops (Dudley et al 1989).

Institutional considerations

Current international input

The present disposition of entomologists concerned with pests of legumes employed by international agricultural research centres in Asia is: Asian Vegetable Research & Development Centre (AVRDC), Taiwan, one International Scientist; ICARDA, Syria, one International Scientist; ICRISAT, India, one Principal Scientist, one Associate Principal Scientist and one National Scientist.

These institutes receive significant input from specialists from other 'Western' organizations and national programmes. With such a small core group of scientists the need for effective setting of priorities is paramount. There is, for instance, no place for the creation of competition with national programmes.

Several 'Western' agencies and non-governmental organizations also work with Asian national programmes on important research and extension problems related to IPM for legumes. There is no doubt that the knowledge generated by these interactions could be applied beyond the limits of the existing bilateral relations.

My impression is that, outside India (Sachan 1991), there are not many national programme scientists in Asia with responsibility for legumes' pest management, especially when compared to those working on pest management for rice. Those whom I know are dedicated and work hard within the constraints set by their budgets and the demands put on them by international cooperators. They need more support in view of the important task they have.

Future requirements

There are unpleasant connotations in not modifying the status quo. Given my views on why stagnation in legume production exists, I am duty bound to indicate how the situation can be improved. How can more legumes be made available to the people of Asia?

(1) Governments can import them, and pay for them by persuading their farmers to grow cash crops that enter international trade. This route to food legume security has a large pitfall: *these products or an acceptable substitute still have to be grown by a pest-confounded farmer somewhere.*

(2) Empower Asian farmers to increase their productivity by adopting IPM. A general policy and constraint list is presented above, but IPM is a highly knowledge-intensive activity. A decision to put high priority on the need to collect and disseminate this information and to follow through with the appropriate action is needed at the international level.

What is needed is a pivotal (= Asian) research and training clearing house for legume IPM matters. This could be part of an existing institute (an International Agricultural Research Centre or a centrally placed National Programme or University), part of a regional agricultural research centre or of an Asian legumes research institute (if one of these attractive concepts should one day exist). It could also operate as a stand-alone 'International Plant Protection Institute'.

Irrespective of the affiliation, the functions of such an organization would be clear: to provide the facilities and infrastructure (including access to cooperating farmers, extension workers and research specialists) needed by plant protectionists from developed and developing countries.

The component roles and insect-related activities could include:

(1) Training scientists, technicians and extension officers in all aspects of legume pest management.
(2) Promoting multidisciplinary approaches to solving pest problems.
(3) Linking with and supporting (if necessary) other IPM projects in Asia, e.g. FAO intercountry rice project.
(4) Providing library and literature search facilities.
(5) Preparing and distributing documentation written to meet the IPM requirements of the region.
(6) Fulfilling the regional need for a central insect collection and identification centre (Wightman 1988).
(7) Screening, banking and distributing legume germplasm known to have resistance to pests.
(8) Providing facilities and an infrastructure for breeders to undertake hybridization and multilocational trials.
(9) Providing the high-grade laboratory facilities needed for determining the mechanisms of host plant resistance, immunoassay and virus detection and description.

(10) Maintaining pest cultures for experimental purposes.

(11) Initiating a pesticide resistance monitoring laboratory.

(12) Maintaining a botanical garden of plants of the region that have anti-pest properties.

(13) Plotting pest outbreak and danger areas and carrying out other Geographical Information System operations.

(14) Developing and providing software for statistical analysis, instructional and analytical models, and real-time, farmer-support (expert system) programs.

Conclusions

There is a pressing need to address the failure of the pan-Asian farming sector to increase legume production and productivity beyond the levels of the 1960s. The current trend of increased insect pest attack associated with the overuse of insecticides can only lead to further deterioration in an already grim scene. It is hoped that this analysis of the problem and the short-term and long-term solutions that are presented will lead to further consideration being given to the development of IPM schemes that are suited to the needs of the millions of smallholder farmers. This is an information-intensive process that requires coordination and cooperation in many directions. The development of an organization with a pan-Asian responsibility to ensure that this happens is fundamental in ensuring an up-turn in legume production.

Acknowledgements

It is a pleasure to record my appreciation of the hospitality and intellectual stimulus given to me by Marcos Kogan, Myron Schenk and Alan Cooper of the Integrated Plant Protection Center (IPPC), Oregon State University during the preparation of this paper. I also offer my thanks to the members of the IPPC and ICRISAT colleagues for improving the readability of this paper.

References

Armes NJ, Jadav DR, Bond GS, King ABS 1992 Insecticide resistance in *Helicoverpa armigera* in South India. Pestic Sci 34:355–364

Bidinger PD, Bhavani N 1980 The role of pigeonpeas in village diets. In: Proceedings of the international workshop on pigeonpeas. ICRISAT, Patancheru, India, p 357–364

Brader L 1979 Integrated pest control in the developing world. Annu Rev Entomol 24:225–254

Broto H, Bottema JWT 1990 CGPRT crops in Indonesia: 1960–1990. A statistical profile. CGPRT Centre, Bogor, Indonesia (Work Pap 4)

Byth DE 1980 Critique and analysis—breeding. In: Proceedings of the international workshop on pigeonpeas. ICRISAT, Patancheru, India, p 487–495

Dudley NJ, Mueller RAE, Wightman JA 1989 Application of dynamic programming for guiding IPM on groundnut leafminer in India. Crop Prot 8:349–357

Escalada MM, Heong KL 1993 Communication and implementation of change in crop protection. In: Crop protection and sustainable agriculture. Wiley, Chichester (Ciba Found Symp 177) p 191–207

Forrester NW 1990 Designing, implementing and servicing an insecticide resistance management strategy. Pestic Sci 28:167–179

Hong NX, Nam NH, Yen NT, Tuong LK 1992 First survey of pigeonpea insect pests in Vietnam. Int Pigeonpea Newsl 15:30–31

Hulse JH 1990 Nature, composition and utilization of grain legumes. In: Use of tropical grain legumes: proceedings of a consultants meeting, ICRISAT Center, 27–30 March 1989. ICRISAT, Patancheru, India, p 11–27

ICRISAT 1989 Use of tropical grain legumes: proceedings of a consultants meeting, ICRISAT Center, 27–30 March 1989. ICRISAT, Patancheru, India

Jayaraj S, Rabindra RJ 1993 The local view on the role of plant protection in sustainable agriculture in India. In: Crop protection and sustainable agriculture. Wiley, Chichester (Ciba Found Symp 177) p 168–184

Kenmore PE, Litsinger JA, Bandong JP, Santiago AC, Salac MM 1987 Philippine rice farmers and insecticides: thirty years of growing dependency and new options for change. In: Tait J, Napometh B (eds) Management of pests and pesticides: farmers' perception and practices. Westview Press, Boulder, p 98–108

Legumes Program, ICRISAT 1991 Summary proceedings and recommendations of a workshop on integrated pest management and insecticide resistance management (IPM/IRM) in legume crops in Asia, 19–22 March, Chiang Mai, Thailand. ICRISAT, Patancheru, India

Marwoto, Neering KE 1989 Mungbean pest management by adaptation of planting date and supervised control. In: Proceedings of the first Asia-Pacific conference of entomology, 8–13 November 1989, Chiang Mai, Thailand, p 381–388

Maurya DM, Bottrall A, Farrington J 1988 Improved livelihoods, genetic diversity and farmer participation: a strategy for rice breeding in rainfed areas of India. Exp Agric 24:311–320

Nagarajan S 1993 Plant diseases in India and their control. In: Crop protection and sustainable agriculture. Wiley, Chichester (Ciba Found Symp 177) p 208–227

Purseglove JW 1968 Tropical crops: dicotyledons. Longman, Essex

Sachan JN 1991 IPM of pulse pests in India. In: Summary proceedings and recommendations of a workshop on integrated pest management and insecticide resistance management (IPM/IRM) in legume crops in Asia, 19–22 March, Chiang Mai, Thailand. ICRISAT, Patancheru, India, p 17–18

Smartt J 1976 Tropical pulses. Longman, Essex

van Emden HF 1980 Critique and analysis—entomology. In: Proceedings of the international workshop on pigeonpeas. ICRISAT, Patancheru, India, p 477–480

Varma A 1993 Integrated management of plant viral diseases. In: Crop protection and sustainable agriculture. Wiley, Chichester (Ciba Found Symp 177) p 140–157

Wallis ES, Woolcock RF, Byth DE 1988 Potential for pigeonpea in Thailand, Indonesia, and Burma. CGPRT, Bogor, Indonesia

Wightman JA 1988 Some solutions to insect-identification problems. Int Arachis Newsl 3:18–20

Wightman JA 1989 The status of insect pests in Asian legume crops 10 years from now. In: Proceedings of the first Asia-Pacific conference of entomology, 8–13 November 1989, Chiang Mai, Thailand, p 394–405

Wightman JA, Ranga Rao GV 1993 Groundnut insects. In: Smartt J (ed) The groundnut crop. Chapman & Hall, London, in press

Wightman JA, Anders MM, Rameshwar Rao V, Mohan S 1993 The management of *Helicoverpa armigera* (Lepidoptera: Noctuidae) on chickpea in southern India: thresholds and the economics of host plant resistance and insecticide application. Crop Prot, in press

Wightman WR, 1990 Farming systems and water management in an Alfisol watershed in the semi-arid tropics of southern India. Vauk Science Publishers, Kiel (Farm Syst Resour Econ Trop 6)

Zadoks JC 1991 A hundred and more years of plant protection in the Netherlands. Neth J Plant Pathol 97:3–24

DISCUSSION

Neuenschwander: The important issues in legume production are to maintain adequate yields and to reduce dependence on pesticides. West African farmers get yields of about 300 kg ha^{-1} for cowpea. This yield is sustainable and there are not many problems with pests. My colleagues tell me that they can't increase yield unless they use other varieties. If they use other varieties, they have to spray them. The amount sprayed has been reduced considerably from what was used, let's say 10 years ago, but cowpea still needs three well-timed sprays, which for West African smallholders is very difficult.

To raise yields without using insecticides, we have to become much more enterprising, for instance in biological control. We have to jump even the continental perspective and become global: we have to try biological control by importing insects from different pulse crops from different continents. Cowpea is indigenous to West Africa and all our pests are supposed to be indigenous, although no serious attempt has been made to investigate their true origin. Some probably originated outside West Africa, which raises possibilities for biological control. We have some indications that this would work, but much more international collaboration is needed.

For such collaboration, the International Institute of Biological Control of CABI in London is trying to establish regional centres of biosystematics. The first is in Central America; IITA will become another regional centre of this group, which is referred to as BIONET. There is a conference in June 1993 on establishing this network of collaborating institutions. It is not specially for cowpea, but pulses are one area where the need is most felt.

Wightman: Earlier, we have discussed attainable yield and potential yield (Rabbinge, this volume). In Asia, current yields of legumes are about the same as in Africa, 300–500, perhaps 700 kg ha^{-1}. The yield potential is 3–5 t ha^{-1}, even 8 t ha^{-1}. The attainable yield is about half that, 2–4 t ha^{-1} in farmers' fields. This is the bright point: we can look towards a fivefold increase in yield. If this were not theoretically attainable, there would be no point in seeking increases in productivity by improved pest management.

Not all crops are affected by pests in the same way. Defoliators have little effect on pigeonpea and groundnut production. The sequence of attack is also important. For pigeonpea we can expect perhaps 90% destruction of seed by insects: but that's 90% of the seed which is left after the diseases have taken their toll.

You mentioned biological control and approaches to pest management. I am worried that as insects move around the world their genetic structure changes. What may be appropriate in one place simply may not work in other places. A prime example is the thrips vector of bud necrosis disease and tomato spotted wilt virus. There is tremendous variation in what happens in one alleged species and another. *Maruca* is pandemic; we have to find out whether *Maruca* in West Africa is the same species as *Maruca* in, say, Sri Lanka.

If we start at the position where a pest is a pest because of the overuse of insecticides, we have to find a substitute for insecticides, to wean farmers from their chemical prop. The first approach should be a search for varieties with more pest resistance than the traditional ones have. Once we have resistant varieties, the pest situation will stabilize a little. We can then look at farming systems and ways of improving natural control processes. This should again reduce insecticide use considerably. Although it is possible to apply insecticides within these systems as part of an IPM procedure, the goal is usually to reduce this component to almost zero. Once we have reduced insecticide use, we can look at conventional biological control—the introduction of new organisms. That is a viable sequence, but it doesn't prevent us trying to look at the possible endpoint, the enhancement of natural control procedures, from the very beginning. The research phase of biological control projects is pretty lengthy, after all.

Rabbinge: You mentioned the stagnation in the productivity of legume crops, especially pulses. This phenomenon occurs not only in Asia, but also in many places in Europe. For that reason, about 15 years ago, a Concerted Action was started in different countries in Europe to determine the reasons for stagnation of yield increases in legumes, especially in faba bean. There was also great variability in yields: the stability of yields in many places in Europe was very low. The European Community wanted to promote the production of faba bean as a replacement for imports of soybean from the USA. It was thought that the low yields were due to pests and diseases. However, investigations showed that it was not only the yield-reducing factors that were important, yield-limiting factors were too. Water and phosphorus are limiting during part of the growing season. If proper water management is achieved (that means water should not always be readily available; at some time during crop development, especially around flowering, there is a need for some stress, a little water shortage), one gets very high yields and stable yields. When nutrient management is not correct, pests and diseases simply amplify the instability and the yield depression. So if you are starting such a programme, it is important to consider not only the yield-reducing factors but also the yield-limiting factors.

Wightman: We are aware that the abiotic factors are extremely important. We even have quite a lot of evidence, which isn't always accepted, that the effects of pests on pigeonpea and groundnut are made worse by water stress.

Nagarajan: Industrialists in Bombay are considering manufacturing a pulse analogue, a synthetic pulse, similar to 'nugget' from soybean, which tastes like meat when cooked. Is ICRISAT considering any research on a pulse analogue for value addition to millet flour, for example?

Secondly, the extent of genetic diversity in pigeonpea is very narrow, primarily because it's a crop of the subcontinent. By using biotechnological tools, we should be able to generate more diversity in pigeonpea.

Wightman: I'm sorry I don't know anything about the pulse analogues.

The biodiversity of pigeonpea is fairly narrow, but there are several pest-resistant genotypes within the species. We have considered very carefully whether we should attempt to amplify the genetic resources of pigeonpea with specific genes from *Bacillus thuringiensis* (Bt) to make selected varieties resistant to *Helicoverpa*. So far, we have not used this approach, simply because we don't have the right Bt strains. In laboratory conditions, a species closely related to *Helicoverpa armigera* has become resistant to Bt in 27 generations. If we take five years to develop a suitable plant with the Bt gene in it, and it takes 27 generations (three years) for the resistance to break down, that is not a good investment. But there is much work going on with Bt and new strains are steadily becoming available; so I regard this as being in abeyance rather than forgotten. There are other potential resistance mechanisms that may be tried in due course. Incidentally, I have been encouraged by what I've seen in the US. They have potatoes carrying Bt genes that are virtually immune to Colorado beetle, which is excellent.

Kenmore: I support your reluctance to use Bt genes for resistance to *Helicoverpa*. For 3–4 years, people from industry and some short-sighted academics have been saying 'All we have to do is switch in the DNA sequence encoding the receptor site like a cassette and we can take care of the resistance problems'. Gould et al (1992) have developed a strain of *Helicoverpa* in which the resistance to Bt is not dependent on binding site specificity. Darwin is still right: do not put Bt genes into crop plants. Make more virulent Bt that farmers can use to hit a pest population once when it's needed—that's fine. Make Bt more specific to different targets and more lasting in an environmental condition, but do not put it into the crop plant.

Nagarajan: What's the pest scenario with regard to the new hybrid pigeonpea?

Wightman: This is really a kind of dwarf pigeonpea. It is highly susceptible to insect pests; I don't think it can be grown without support from insecticides. As such, it cannot be part of a sustainable farming system.

Swaminathan: I agree with Dr Wightman about the inadequate research effort in legumes in the past. As far as varietal improvement is concerned, it is not just a lack of adequate scientific or breeding effort. There are inherent difficulties in combining yield and quality. The last 50 years work from various laboratories around the world trying to combine high protein yield with high calorie yield has shown that whenever we make a gain in calorie yield, there is a protein penalty. Whenever there is a protein yield improvement, there is a calorie penalty. We have to find a compromise between the two.

But there are crops, like peas in Europe and North America, for which very high yields have been obtained. Soybean is another case. In India, soybean is a new crop. In Tamil Nadu, for example, many farmers produce $4-5\,t\,ha^{-1}$ of soybean, largely because they started with a completely new set of management procedures. They didn't have their own past memory as to how this crop should be cultivated. Similarly, there are reports from the CSIRO in Australia that some of the pigeonpea varieties taken from India have given very high yields, probably under conditions of good management in terms of plant population and plant protection. So we have to be realistic about what can be achieved through breeding alone or biotechnology; we should also consider crop management.

Varma: Your list of insect pests (Table 3) did not mention *Madurasia obscurella*, a beetle which is very common in India. Is it not a pest of grain legumes? Also, don't thrips cause some direct damage to grain legumes? You mentioned them only as a viral vector.

Wightman: The insects I listed are pan-Asian pests; the beetle you mentioned is rather more localized. Mung bean can be wiped out by thrips; in groundnut they cause only cosmetic damage, except as virus vectors.

Varma: For storage pests in pulses, people use vegetable oils, like groundnut oil, mustard oil, palm kernel oil, etc. These are very effective in preventing damage by beetles (Pereira 1983).

Wightman: Dusts and oils can be used to protect pulses from storage pests. This is rediscovered about every three months, if you look at the literature! Unfortunately, the message does not always get through to the farmers. There are still farmers who are losing pulse seed. This is one reason they don't keep their own seed—they think they are going to lose it to insects and moulds.

Jayaraj: In pulse production we need to have a policy decision in many developing countries, based on the successes and failures of the past. In pigeonpea, varietal improvement has not really made an impression, compared to other crops. As Dr Swaminathan mentioned, management is very weak in pigeonpea and in chickpea. For groundnut, the yield breakthrough has been remarkable. For soybean, cowpea, green gram (mung bean) and black gram, increased yields are being obtained. In peninsular India, where mung bean and black gram are grown after the rice crop, the yields are picking up. There are not many pest problems because of improved varieties. Yellow mosaic virus used to be a problem but there are now effective resistant varieties.

But in the case of both pigeonpea and chickpea, we have come to a dead end. Of all the grain legumes, the protein quality of these two pulses seems to be the poorest, though by food habits we prefer these two pulses in many Asian countries, particularly in India. The time has come to change, if possible, our eating habits, because of this dead end in improving pigeonpea and chickpea productivity. Or we should have a very strong programme for managing the pests and diseases, which compete with humans for protein in the form of grain legumes. Plant stand in the field is badly affected. Damage of pulses by beetles during storage reduces germination and then in the early stages of cultivation half a dozen pests cause extensive damage. Often only one quarter or one fifth of the field is covered with standing plants and subsequently this is affected by many viral diseases and so on. In India, *per capita* consumption of pulses has not increased in the last 40 years. So some policy decision is necessary for sustaining production of pulses. If we cannot make scientific advances, we should change the cropping pattern.

Wightman: I don't agree; there is terrific potential still in chickpea. For pigeonpea, the problem is finding an alternative crop that will grow in these harsh environments. Pigeonpea will grow where other crops won't grow.

Lakshmi: There is a genus called *Atylosia* Wt & Arn, which is a wild species. It is resistant to almost all insects and pests. It is a valuable germplasm material and very closely related to *Cajanus* (L.) Mill sp. Professor N.C. Sburamanyam at the University of Hyderabad, India, claims to have obtained a cross between *Atylosia* and pigeonpea (personal communication, 1990, with Professor M. K. Rao, who has seen this hybrid). Is it true that *Atylosia* can be crossed with pigeonpea and resistance genes transferred to pigeonpea?

Wightman: Today, anything is possible. *Atylosia* is a wild species, a very close relative of pigeonpea. Interspecific crosses have been made at ICRISAT, but they have not been successful as far as insect resistance is concerned. If Professor Sburamanyam has done this, we would be very interested to work with him and take it further.

Zadoks: I would like to challenge the plant breeders, represented by Mr Mishra. I think the breeders should breed a variety, hybrid or not, which is useful to farmers. But with the seed, they should also sell a recipe for integrated management of the crop. Is that a feasible proposition?

Mishra: There are All-India Coordinated Improvement projects for different crops. The resistant varieties and hybrids of different crops have been tested under certain sets of conditions. When the farmers adopt those hybrids or varieties, variation in susceptibility is observed. For maintenance of the genetic resistance, correct management practices need to be adopted along with the new varieties.

A pigeonpea hybrid ICPH-8 has been released from ICRISAT, but it is as susceptible to *Fusarium wilt* and Sterility mosaic virus as any variety developed from other programmes. Good hybrids and varieties require good management support to realize their yield potential.

As a seed company, we market seeds of hybrids and improved varieties. A leaflet giving guidelines for the adoption of better agronomic practices is supplied with each individual seed bag.

References

Gould F, Martinez-Ramirez A, Anderson A, Ferre J, Silva FJ, Moar WJ 1992 Broad spectrum resistance to *Bacillus thuringiensis* toxins in *Heliothis virescens*. Proc Natl Acad Sci USA 89:7986–7995

Pereira J 1983 The effectiveness of six vegetable oils as protectants of cowpeas and bambara groundnuts against infestation by *Callosobruchus maculatus* (F.) (Coleoptera: Bruchidae). J Stored Prod Res 19:57–62

Rabbinge R 1993 The ecological background of food production. In: Crop protection and sustainable agriculture. Wiley, Chichester (Ciba Found Symp 177) p 2–29

Perspectives for crop protection in sustainable agriculture

M. S. Swaminathan

Centre for Research on Sustainable Agricultural & Rural Development, 3rd Cross Street, Taramani Institutional Area, Madras, 600 113, India

Abstract. The challenge facing agriculture, particularly in developing countries, is to increase productivity without causing ecological damage. An important aspect of sustainable agriculture is the substitution of chemicals and capital with locally grown biological inputs and knowledge. Production of food grains in India has risen steadily since the 1960s, partly through the introduction of high-yielding varieties of rice and wheat together with appropriate agronomic and plant protection practices. Plant breeding programmes have improved cultivated rice by the transfer, from wild species, of genes conferring resistance to viral diseases. Diversification of resistance is sought, to render the hybrids less vulnerable to sudden outbreaks of diseases and insect pests. For wheat, resistance to rusts involves the additive interaction of several gene-encoded traits. Linkage of resistance genes with morphological markers enables the inheritance of the former to be followed easily. It is important that recent advances in molecular genetics are incorporated into integrated pest management programmes. This will require appropriate social organization. Research centres that develop and exploit genetic sources of resistance and varietal diversification should be established.

1993 Crop protection and sustainable agriculture. Wiley, Chichester (Ciba Foundation Symposium 177) p 257–272

Agenda 21 of the United Nations Conference on Environment Development (UNCED) places emphasis on the linkages between alleviation of poverty and sustainable management of resources. It notes that by the year 2025, 83% of the expected global human population of 8.5 thousand million will be living in developing countries. For the poor in developing countries, agriculture is both a way of life and the primary means of livelihood. Growth in non-farm employment opportunities has been slow and agriculture therefore remains the main avenue for food and livelihood security. According to Agenda 21 of UNCED, 'agriculture has to meet this challenge, mainly by increasing production on land already in use and by avoiding further encroachment on land that is only marginally suitable for cultivation'. How can crop productivity per units of land, water, energy and time be increased continuously, without causing harm to the ecological foundations on which further advances in productivity depend?

It is now agreed that an important component of technologies which can help to promote sustainable agriculture is the substitution of capital and chemicals with knowledge and farm-grown biological inputs. Such substitution will, in addition, help to achieve a reduction in the cost of production without lowering yield. Hence, ecological agriculture is also the method of choice for resource-poor farmers, because it helps them to overcome to some extent their economic handicaps. Mineral fertilizers as sources of nutrients and pesticides designed to meet the challenge of pests, pathogens and weeds constitute two major sources of ecological problems, particularly when they are applied excessively and inappropriately. The efficiency of crop protection is a major determinant of yield in tropical Asia, particularly during the wet (monsoon) seasons, when atmospheric humidity remains high and conditions for pest epidemics are favourable.

Famines were frequent in India during the 19th century. They have been avoided in this country since the 1950s through concurrent attention to enforcing production and consumption. The production of food grains has grown steadily, thanks to the implementation of mutually reinforcing packages of technology, services and public policies (Fig. 1). Much of the progress is due to the intro-duction of high-yielding varieties of wheat and rice along with appropriate agronomic and plant protection practices in the mid-1960s. However, with a change in the ecology and microenvironment of the crop, problems of pests and diseases also increased (Table 1). The pedigrees of new varieties are thus complex, since the breeder attempts to assemble in one genotype genes for resistance to a wide range of pests and diseases.

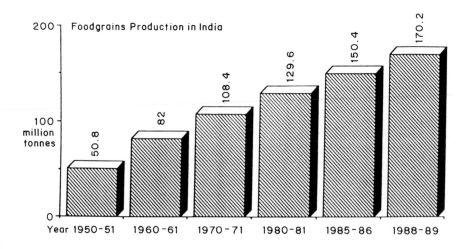

FIG. 1. Production of food grains in India. Data provided by the Ministry of Agriculture, Government of India.

TABLE 1   The changing pest scenario in India with the emergence of new pests and biotypes

1965	1975	1980	1985	1990
Stem borer	Stem borer	Stem borer	Stem borer	Stem borer
Gall midge	Gall midge	Gall midge	Gall midge	Gall midge (3 biotypes)
Green leaf hopper	Green leaf hopper	Green leaf hopper	Green leaf hopper	Green leaf hopper
	Brown plant hopper	Brown plant hopper	Brown plant hopper	Brown plant hopper
	White-backed plant hopper	White-backed plant hopper	White-backed plant hopper	White-backed plant hopper
		Cut worm	Cut worm	Cut worm
		Gundhibug	Gundhibug	Gundhibug
		Hispa	Hispa	Hispa
			Leaf folder	Leaf folder
				Thrips
				Mites

Rice and wheat are the principal anchors of the food security system of India. Rice is grown during the wet and dry seasons, while wheat is grown only during the winter season (November to April), although small pockets of wheat cultivation occur in the hills during summer. Plant protection for sustainable agriculture involves in both cases integrated pest management (IPM) procedures. Genetic resistance and social mobilization are important for the success of IPM procedures on small holdings. Hence, only these two aspects are dealt with in this paper.

Crop protection in rice

Since 1930, substantial progress in bridging the gap between actual and potential yields has been made in rice grown on experimental stations (Fig. 2). Such progress has come through changes in plant architecture, leading to a better response to water and fertilizer, and through the exploitation of hybrid vigour. The challenge now is to impart stability to yield through resistance/tolerance to pests and diseases. This is particularly important in rice because, next to cotton, rice cultivation accounts for the greatest use of pesticides.

Wild species of *Oryza* are a rich source of useful genes for the improvement of cultivated rice. Nevertheless, efforts to introduce useful traits from wild species to cultivated rice have been limited. Some successful examples are the transfer of a gene for grassy stunt virus resistance from *O. nivara* and transfer of cytoplasm of *O. perennis* to obtain cytosterile lines for use in hybrid rice

Potential rice yield (t/ha)

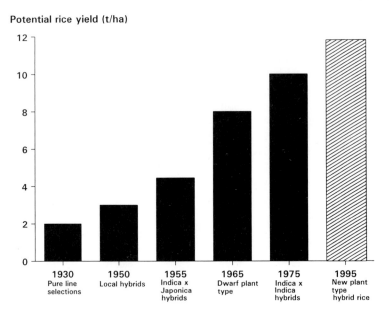

FIG. 2.   Successive increases in yields achieved with new varieties of rice.

breeding (see IRRI 1985). In recent years, disease and insect outbreaks on rice have become more intense and germplasm resources of rice, including wild species, have been evaluated to identify donors for resistance. Several wild species and their accessions have been identified to be resistant to all known biotypes of brown plant hopper, white-backed plant hopper and green leaf hopper (IRRI 1985). In interspecific hybrids with *O. officinalis* and *O. sativa*, it was possible to recover lines resistant to the brown and white-backed plant hoppers, comparable in yield to the best *O. sativa* parents. *O. minuta*, a tetraploid species native to Asia with genomic composition of BBCC, is a potential source of resistance to two important rice diseases—bacterial leaf blight and rice blast. Two accessions of *O. minuta* that had a high level of resistance against isolates of six races of *Xanthomonus campestris* pv. *oryzae* from the Philippines and blast isolate PO-66 were hybridized at the International Rice Research Institute with susceptible *O. sativa*. The hybrids with *O. minuta* accession 101141 were resistant to all *X. campestris* isolates. Work is in progress to confirm whether the alien resistance is non-allelic to known sources of resistance or not (Khush 1989).

Diversification of cytoplasmic male sterility

More than 95% of the male sterile lines used in *indica* hybrids in China belong to the *WA* (*Wild Aborted*) system. The same is true of a large number of improved cytoplasmic male sterility lines now being experimented with in India and elsewhere. There is a continuous search for alternate sources of cytoplasmic male sterility, because dependence on a single source for such an important trait would render the hybrids vulnerable to sudden outbreaks of diseases and insect pests. Research efforts are underway in India to diversify the sterility system by looking for new sources from both cultivars and wild species of the genus *Oryza*, as well as by searching for restorer genes (i.e. those able to restore fertility) for already developed male sterile lines. Research efforts during recent years have helped the Directorate of Rice Research at Hyderabad, India to suspect the presence of exploitable male sterility in 1–2 accessions each of five species of A-genome—*O. glaberrima, O. nivara, O. rufipogon, O. barthii* and *O. longistaminata*. With one backross using *O. sativa* as the recurrent parent, achievable sterility has gone up to 100% (Table 2) (E. A. Siddiq, personal communication).

MS 577 and IR-64A are as stable and have the same potential as male sterile lines as *WA*-based cytoplasmic male sterility lines. They have not been used owing to a lack of effective restorers. Because they were suspected to have originated from closely related wild species, they were crossed with a large number of Indian and exotic accessions of A-genome species. None of the accessions tested was an effective restorer, but a few genotypes capable of partial restoration were identified (Table 3). Possibilities for improving the level of fertility restoration

TABLE 2   Evaluation of interspecific crosses for alternate sources of sterile cytoplasm in *Oryza*

Cross	F1		Backcross 1	
	Pollen sterility (%)	*Spikelet fertility* (%)	*Pollen sterility* (%)	*Spikelet fertility* (%)
O. nivara/O. sativa (Acc. 1-1) (IR-70)	86	5	95	2.5
O. nivara/O. sativa (Acc. 1-2) (IR-64)	90	5	100	0.0
O. rufipogon/O. sativa (Acc. 2-1) (IR-70)	70	10	90	3.5
O. rufipogon/O. sativa (Acc. 2-2) (IR-64)	100	0	100	0.0
O. rufipogon/O. sativa (Acc. 2-3) (IR-64)	100	0	100	0.0
O. longistaminata/O. sativa (Acc. 3-1) (IR-70)	100	0	—	—
O. barthii/O. sativa (Acc. 4-1) (IR-58025B)	100	0	—	—

Source: Directorate of Rice Research, Hyderabad, India (Dr. E. A. Siddiq, personal communication).

TABLE 3   Identification of restorer sources for MS 577, a rice line carrying cytoplasmic male sterility

	Fertility (F1)	
Cross	Pollen (%)	Spikelet (%)
Pushpa A/*O. rufipogon* (Acc. 3)	80	50
Pushpa A/*O. rufipogon* (Acc. 4)	50	30
Mangala A/*O. rufipogon* (Acc. 5)	30	20

Source: Directorate of Rice Research, Hyderabad, India (Dr. E. A. Siddiq, personal communication).

by intercrossing among the partial restorers of wild origin, as well as with single and two gene-governed restorers of *WA* source, are being explored. Diversification of the source of cytoplasmic sterility will help impart greater stability of resistance mechanisms.

The Rice Biotechnology programme supported by the Rockefeller Foundation has led to standardization of techniques for genetic enhancement through novel genetic combinations. This approach is likely to prove useful in the development of high-yielding strains possessing resistance or tolerance to a wide range of biotic and abiotic stresses. A recent book on *Rice biotechnology* edited by G. S. Khush and G. H. Toenniessen (1991) contains an excellent summary of advances in the application of the tools of molecular biology in the breeeding of disease- and pest-resistant rice varieties.

Crop protection in wheat

Historical accounts of rust epidemics and deployment of resistance genes in certain geographic areas have provided information on the host–parasite interaction (Watson & Luig 1963). It is clear that certain gene interactions are more durable than others (Van der Plank 1963). The value of partial resistance was known in the last century (Farrer 1898). Fifty years of deployment of hypersensitive genes have produced a periodic boom and bust cycle. This has intensified the search for durable resistance.

Breeding for slow rusting/partial resistance: a step towards durability

Slow rusting (Caldwell 1968) is when the rate of disease development is delayed compared to that in susceptible cultivars (Rajaram et al 1988) and the final disease severity is low. This kind of host–parasite interaction is characterized by a susceptible reaction. Several plant characters may interact to produce a high level of partial resistance. In wheat, partial resistance (slow rusting) to rusts has been found to involve exclusion and low receptivity to the fungus

(Romig & Caldwell 1964, Sartori et al 1978), reduction in uredial size, and a longer latent period (Parlevliet 1979).

Identification of components of partial resistance

Longer latent period, smaller uredinia size and fewer uredinia per unit area play strong roles in retarding disease development in the field. Singh et al (1991) quantitatively evaluated adult plants of 28 CIMMYT International Maize and Wheat Research Center, Mexico, bread wheat varieties with a pathotype of *Puccinia recondita* f. sp. *tritici*. They studied the components of partial resistance. Though all varieties displayed high reaction (severe infection), an increase of 14 to 49% in latent period, a decrease of 42 to 98% in uredial number, and a decrease of 34 to 78% in uredial size were noted compared to the susceptible cultivar Morocco. The area under the disease progress curve in the field for these varieties was 50 to 99% less compared to Morocco. Significantly, high negative correlations between latent period and uredial number, uredial size and disease progression, and positive correlations between uredial number and uredial size led to the postulation that the components of partial resistance were either under pleiotropic genetic control or due to closely linked genes. Das (1990) confirmed this hypothesis by studying derivatives of crosses involving a partially resistant and a susceptible cultivar. These studies clearly indicated that partial resistance could be easily accumulated by selecting for plants or lines that show low rust severities compared to the susceptible checks (varieties with high susceptibility which serve as indicators of the efficacy of the screening procedures). Genetic diversity was also evident.

Durable resistances to rust diseases and their linkages with morphological markers

The *Sr2* complex derived from the variety Hope and the related line H44-24 seems to have provided the foundation for durable resistance to stem rust in CIMMYT and CIMMYT-derived semidwarf germplasm. Gene *Sr2* confers a slow rusting response when present alone, which occasionally is not acceptable. However, the *Sr2* complex, defined here as the additive interaction of *Sr2* with other known or unknown slow rusting genes, results in negligible final severity of stem rust. *Sr2* is known to be linked to gene *Pbc*, which confers the pseudo black chaff phenotype. *Pbc* is often used to check the presence of *Sr2* in the presence of other race-specific or race-unspecific minor genes.

The *Lr34* complex derived from the Brazilian cultivar Frontana has been judged as one of the best sources of durable resistance to leaf rust (Roelfs 1988). Singh & Rajaram (1992a,b) found that Frontana and some other CIMMYT wheats displaying very low final disease severity carry *Lr34* and 2–3 additional genes. Genetic analysis of numerous wheats which display various levels of partial resistance suggests that such resistance is controlled by the interactions of relatively few genes. A combination of 3–4 slow rusting genes in a cultivar results in a very low level of disease in all environments. Singh (1992a) has confirmed the genetic linkage between *Lr34* and gene *Ltn*, which causes leaf tip necrosis of adult plants. *Lr34* and *Ltn* occur in approximately 60% of current CIMMYT germplasm and were postulated to be present in several Indian wheats (Singh & Gupta 1991). Recent genetic studies at CIMMYT (R. P. Singh, unpublished) have indicated that variability for the slow rusting genes is high. For example, Pavon 76 and Genaro 81 each carry two different additive genes, which are different from those in Frontana and some other *Lr34*-carrying wheats. All these cultivars have demonstrated the durability of leaf rust resistance.

Stripe rust resistance due to the Yr18 complex

Singh (1992b) reported that *Lr34* was linked to gene *Yr18*, which confers a slow rusting type of adult plant resistance to stripe rust. The durable resistance of Anza is due to *Yr18*. As with *Sr2* or *Lr34*, other slow rusting genes were found that interact additively with *Yr18* (R. P. Singh, unpublished). An accumulation of two or three such genes with *Yr18* results in very low final stripe rust severity. Leaf tip necrosis can again be used to monitor the presence of *Yr18*. Variability for additive genes effective in adult plants is also high.

Recent research on the three rust diseases indicates that durability can be achieved by utilizing combinations of additive slow rusting genes. Diversity can be maintained around the *Sr2*, *Lr34* or *Yr18* genes and the presence of these genes can be followed by morphological markers. Thus, new technologies and products are likely to help in imparting greater stability to the resistance of crops to pests and diseases. There is, however, a need to ensure that IPM programmes benefit from recent advances in molecular genetics. There are differences of opinion with reference to sustainable benefits from such technologies. For example, the use of genes from *Bacillus thuringiensis* for pest control is advocated in cotton as a means of reducing the use of pesticides. There is, however, a fear that their widespread use may lead to the build-up of resistance in insect populations. In order not to lose the confidence of farmers in new techniques, it is essential that we analyse both potential benefits and potential hazards carefully before finalizing extension recommendations. Also, there is need for the establishment of research centres where appropriate biological software (such as donors of resistance to pests) can be assembled for use in the breeding of location-specific varieties. Biological software libraries for achieving durable resistance and varietal diversification will be of much help to breeders and adaptive research workers at the 'grassroots' level.

A farming community level approach to IPM

Experience in many countries reveals that the availability of an effective IPM technology alone is no guarantee that it will prove effective in the field. A top-down technology-driven approach will not succeed, particularly where the average size of a farm is one hectare or less and farms, in addition to being small, are also fragmented in their spatial distribution. The development of IPM technologies under such conditions should involve the active participation of farm families, particularly women who play a dominant role in both seed selection and crop management.

Various forms of social mobilization are now being tested for promoting group efforts in pest management. Under the Biovillage programme of the Centre for Research on Sustainable Agricultural and Rural Development, Madras, village youths are being trained and assisted in undertaking pest proofing of the entire village. If all the farmers with smallholdings entrust the responsibility for pest management to a village group of trained women and men, the adoption of integrated pest management procedures will be easy. Inadequate attention to the promotion of group efforts has resulted in the non-adoption of the tools of IPM. Group extension and group insurance and other public policy instruments will also help.

There are real opportunities today for realizing a high proportion of the potential yields of crops on an ecologically sustainable basis. This will, however, require greater interaction between plant protection experts and social scientists. Also, it will be necessary to establish biological software centres for sustainable agriculture.

Summary

In many developing countries, particularly in South and South-East Asia, there is little scope for increasing food production through area expansion, except through a higher cropping intensity in irrigated areas. Much of the needed increase will have to come from an increase in productivity per units of land, water, time and energy. Multiple cropping and the occurrence of alternate hosts promote the greater incidence of insect pests, pathogens and weeds. The problem is particularly acute during the wet season. Ecologically sound pest management thus holds the key to sustainable food security in this region.

Advances in molecular biology and genetic engineering have opened up new opportunities for the accumulation of genes in single varieties through hybridization of distantly related strains. Care should, however, be taken that the strains developed through recombinant DNA experiments will not be susceptible to any early breakdown of resistance. This will be essential for building the confidence of farmers in crop varieties that derive their resistance from novel genetic combinations. Research on the development of biological software for sustainable agriculture needs stepping up, since agricultural progress will hereafter be driven largely by biological findings.

India has nearly 90 million operational holdings, while China has about 100 million. Under such conditions, integrated pest management strategies require for their success appropriate social organization. New methods of protecting entire villages from pests through service centres operated by trained village youth are needed. Unless there is a match between technology and social organization, integrated pest management methods, so very essential for sustainable agriculture, will not be effective.

References

Caldwell RM 1968 Breeding for general and/or specific plant disease resistance. In: Shepherd KW (ed) Proceedings of the 3rd International Wheat Genetics Symposium, Canberra, CSIRO, ACT, p 263–372

Das MK 1990 Inheritance of durable type disease resistance to leaf rust in wheat (*Triticum aestivum* L. em Thell). PhD thesis, Oregon State University, Oregon, OR

Farrer W 1898 The making and improvement of wheats for Australian conditions. Agric Gaz NSW 9:131–168

IRRI 1985 International rice research: 25 years of Partreshy. International Rice Research Institute, Philippines

Khush GS 1989 Multiple disease and pest resistance for increased yield stability in rice. In: Progress in irrigated rice research. International Rice Research Institute, Philippines, p 79–92

Khush GS, Toenniessen GH (eds) 1991 Rice biotechnology. CAB International, Wallingford (Biotechnol Agric Ser 6)

Parlevliet JE 1979 Components of resistance that determine the rate of epidemic development. Annu Rev Phytopathol 17:203–222

Rajaram S, Singh RP, Torres E 1988 Current CIMMYT approaches in breeding wheat for rust resistance. In: Simmonds NW, Rajaram S (eds) Breeding strategies for resistance to the rusts of wheat. CIMMYT, Mexico, p 101–118

Roelfs AP 1988 Resistance to leaf and stem rusts in wheat. In: Simmonds NW, Rajaram S (eds) Breeding strategies for resistance to the rusts of wheat. CIMMYT, Mexico, p 10–22

Romig RW, Caldwell RM 1964 Stomatal exclusion of *Puccinia recondita* by wheat peduncles and sheets. Phytopathology 54:214–218

Sartori JF, Rajaram S, Bauer ML 1978 Bases patologicas y geneticas relacionadas con la resistencia general del trigo a *Puccinia graminis* Pers. F. sp. *tritici* Eriks et Henn. Agrociencia 34:3–16

Singh RP 1992a Association between gene Lr34 for leaf rust resistance and leaf tip necrosis in wheat. Crop Sci 32:874–878

Singh RP 1992b Genetic association of leaf rust gene Lr34 with adult plant resistance to stripe rust in bread wheat. Phytopathology 82:835–838

Singh RP, Gupta AK 1991 Genes for leaf rust resistance in Indian and Pakistani wheats tested with Mexican pathotypes of *Puccinia recondita* F. sp. *tritici*. Euphytica 57:27–36

Singh RP, Rajaram S 1992a Genetics of adult-plant resistance to leaf rust in 'frontana' and three CIMMYT wheats. Genome 35:24–31

Singh RP, Rajaram S 1992b Durable resistance to *Puccinia recondita tritici* in CIMMYT bread wheats: genetic basis and breeding approaches. Vortr Pflanzenzuecht 24:239–241

Singh RP, Payne TS, Rajaram S 1991 Characterization of variability and relationships among components of partial resistance to leaf rust in CIMMYT bread wheats. Theor Appl Genet 82:674–680

Van der Plank JE 1963 Plant diseases; epidemics and control. Academic Press, New York

Watson IA, Luig NH 1963 The classification of *Puccinia graminis* var. *tritici* in relation to breeding resistant varieties. Proc Linn Soc NSW 88:235–258

DISCUSSION

Kenmore: In 1979 at IRRI, before you arrived, Grace Goodell sent a shipment of rice paddy seed taken from a highway in the Philippines to the Commonwealth Mycological Institute for analysis. They found about half a dozen very dangerous mycotoxin-producing fungi in it. So paddy definitely can contain dangerous liver-damaging and other kinds of damaging mycotoxins.

If you go towards increasing the productivity of land through land-saving efforts in other crops as well as rice, from an entomological viewpoint it's extremely important to look at the role of general predators. Those are the ones that will carry over from crop to crop. The specific parasitoids have often been the focus of research in the past because they were easy to look at, and perhaps more important in the temperate zone. The general predators—spiders, beetles, organisms that eat anything that moves—are the things that you can conserve and carry crop to crop.

A more general point is that we have to distinguish carefully between two kinds of approach. The top-down, centrally organized group that was popular in the 1970s doesn't work. In Indonesia, we now have more than 5000 farmers' groups working with IPM. When we did our initial surveys of the pre-existing farmers' groups in Indonesia, fewer than 9% met regularly as directed by government edict. Once you give them something to do, once you get something in the field that benefits the individual farmers both in terms of heightened insight and in terms of mild economic incentives, these groups coalesce very quickly. You can use a strong technological education to strengthen those groups and pull them along instead of pushing them.

In Vietnam, also in China, one reason they are excited about the IPM project is that their top-down groups no longer exist. The communes are now townships; the local village brigades are shattered. They don't know what to do: the government has always been used to working with half a million (commune-sized) units, now they have 150 million (family-sized) units in China. That's when they came to us and said: can we use IPM to get scientific farming moving and coalesce group interest and group support for an idea around that technical concept?

Norton: I would like to look at this from a planning point of view. In land use planning, several people have analysed land suitability looking at the basic resources present in a country, then planning accordingly, to try and maximize biological efficiency. In the past there have been economic plans. We could do similar things for pest control; we could design integrated pest management packages and so on.

The problem with all of these is by the time you have made the plan, the situation has changed. How do we resolve this? There is a Chinese (or is it an Indian?) saying about riding a tiger. We are sitting on a tiger of development: we are trying to guide that development in one way or another. We can't actually design a system and say we are going to change society and that's what it will be like. Do you think we can design systems as targets to work towards, or do we simply get there by incremental means?

Swaminathan: I wish I had an answer. In real life, developments are initiated by the people themselves, then they create their own momentum. But the government in a country like India certainly influences this. The domestic market in India is very large, so, except for plantation crops, we are not strongly influenced by the external market. The Indian success story is not merely due to science; the government has adopted a remunerative pricing policy. Minimum prices are announced at the time of sowing, so they can influence the acreage under various crops. We also have a mechanism to ensure that the minimum price is paid. For example, the Food Corporation of India or a state government agency buys paddy at the announced price, even with a grain moisture content of 20%. This has been a very important stimulant to agricultural growth in India. Even with semi-dwarf varieties of wheat, I don't believe we could have made any progress in terms of production without the support or correct public policy.

But today the problems of development are compounded by the fact that there are minimum needs of the people in terms of clothing, food, shelter and jobs. We have calculated that about 100 million new jobs will be needed in India between now and 2000. Governments have to get re-elected so they look at problems in a very short-term way rather than in terms of sustainability—because sustainability of the party in power is at stake! So we have a very complex situation in terms of policies. The major safeguard in a country like India, is that the country is so large that the government doesn't reach everywhere, which fosters local initiative. We have over 100 000 non-governmental organizations in this country.

I agree with Peter Kenmore about the top-down approach. In our centre, the projects are participatory in nature. We help farmers with knowledge and training, but they raise their own money. Robert Chambers has been pleading for a 'farmer first and last' approach in agricultural sciences. However, what is really effective is an optimal blend of traditional wisdom and experience with modern technology. One should not set one approach against the other; we need a middle road. A 'change in direction' is certainly necessary in agricultural research policies and strategies. We have developed a methodology for imparting a pro-nature, pro-poor and pro-women bias in technology development and dissemination, to foster a new paradigm of rural development based on the integrated application of the principles of ecology, economics and equity.

Kenmore: Even in a much smaller country like Indonesia, which is still the fourth biggest country in the world, there is a great range of policy within which niches can be found for innovative approaches and bottom-up type approaches. There is no question that there must be supportive and protective government policies in order for these local level initiatives to survive. With a large enough scale government, one can usually find them.

The dilemma facing the people in this room is: what should we do? What should scientists do? First, we must have support from government to keep the quality of that scientific work high, so that resistance genes do become available. Then, someone needs to know that they exist and initiate their distribution. But at the same time, we can choose where we are going to work within certain bounds. We have to be able to say we are not going to go for what is intellectually seductive or glamorous in funding terms. We should try and do at least a percentage of our work at village level, doing very location-specific activities. That is where some of these innovations come out, when we can act in partnership with local level initiatives. We must move in that direction.

Waibel: India has a history of subsidizing the agricultural sector. The fertilizer subsidy has just been removed. What impact do you expect that to have on general agricultural development? What alternative use could be made of that money in the context of what Peter Kenmore is proposing? Could you demonstrate to governments that there are much better uses for the money which could be saved on subsidies?

Swaminathan: Subsidies in India have grown. Water is not priced at all; power is cheap. Whenever the fertilizer subsidy has been removed, there has been a reduction in production. In 1981 the subsidy was removed because of recommendations by the International Monetary Fund; immediately, production of food grains fell by about 20 million tonnes. The reason is, the degrees of freedom in terms of improvement of efficiency seems to be low. One important factor in improving the efficiency of fertilizer use is irrigation water management. All the work at IRRI has shown that apart from ammonia volatilization, leaching losses and pest problems decrease the return from the applied fertilizer. Land levelling and consolidation have yet to occur in several parts of India. Today fertilizer use is often a substitute for efficient crop management. So I'm more interested in raising the very low price of water, than in removing the fertilizer subsidy; an improvement in the efficiency of irrigation water management would increase the efficiency of fertilizer use.

Removal of fertilizer subsidy certainly has an immediate detrimental effect on crop production. In the long term, if it improves the efficiency of fertilizer use and promotes the use of organic manure and green manure crops and an integrated nutrient supply system, that will be good. The removal of the subsidy must be accompanied by a great deal of extension work in terms of fertilizer use efficiency and integrated nutrient supply systems. If this happens simultaneously, the adverse effect of removing the subsidy can be avoided.

If production falls next year, there will be an immediate outcry. This year, after several years, the Government of India is importing wheat. This has led to a lot of public criticism. On the other hand, India is in a much better position to avoid famines than are most developing countries because we have over 300 000 'fair price' shops, sometimes called the safety net for the poor. The network is extensive, so even in a year of drought the market price does not go way up. In the Bengal famine 1942–1943, several hundred thousand people died on the roads, while some people were wining and dining in the Great East India hotel. The prices were so high, some could afford them, others could not. Now, the price stability of basic staples in India is remarkable, largely because the government holds large buffer stocks which they can feed into the public distribution systems. So while fertilizer and other subsidies ultimately should go, we should ask why, even in the affluent nations of Europe, farmers are still campaigning for subsidies.

Bhattacharyya: The abolition of the fertilizer subsidy has come very abruptly. The government should have made a plan to increase input efficiency before substitution. The abrupt removal of subsidy might reduce production. There should have been a period of transition, during which the farmers could have prepared for substitution and changed their farming systems. Today, in the villages, the biomass is not available for alternative methods, such as composting.

Rabbinge: This was a nice illustration of how development can be promoted. It is not a question of top down or bottom up, it's always the mixture. A bottom-up approach which is not supported with good facilities through a top-down approach of governments in terms of price subsidies, technology-driven activities and extension, will not be successful. Everyone here is in favour of that type of movement to get development going in a particular direction.

Science can contribute to this development. It can also complement this through devising options for development and by making explicit what options already exist. The prophets of doom who expected a structural famine in the 1980s and in the years to come were fortunately wrong. Science should explore the various options for development and show what ways there are, what potentials there are, and where the limitations are. It should also determine the trade-off between different objectives, say environmental objectives, social objectives, agricultural objectives and also objectives such as proper use of land. By offering such options, you can set goals for which we should aim.

Swaminathan: Variety is the spice of life. One thing is not to have a rigid approach to problems. We need hybrid combinations of technologies. We should distil the best from traditional wisdom and technologies and marry them with the best in modern technology, whether it is biotechnology or information technology, space technology in terms of remote sensing or whatever.

More and more, I find in the literature a polarization. Some people feel everything modern is bad. On the other hand, unless you have a road, you can't take your produce out. The road may require felling a few trees, so there is a trade-off. We don't have enough scientists who are systems analysts, able to look at the problem and calculate the trade-off. In India, we have environmental impact analysts who are often looking at only part of the problem and not the whole. We will have to train more people to look at this problem in a multifaceted way, so that the best option can be clear to both policy makers and the general public.

Norton: I would like to raise an issue which comes from the rice story, possibly from the peanut and groundnut story—that the solution to pest problems is to stop spraying. How do we best get from where we are now to that ideal situation? Obviously, there are transitional problems. If you try to shift to organic agriculture, you have transition problems of various forms. The first is getting rid of the pesticides that are there, because you can't sell your produce for 2–3 years, and farmers therefore have to do without money for a couple of years.

Can we start addressing this problem? Peter Kenmore is doing it in one way in Indonesia. How applicable is that in other cultures? We have to know what things could go wrong and make contingency plans to prevent those happening. Otherwise, as soon as farmers hit a problem, they may revert to using pesticides.

Wightman: I think it's a matter of just doing it! We have to give something as an incentive, such as an improved variety which may have resistance to the pest. For groundnut, we can do this. It is necessary to interest key farmers so they can demonstrate to their neighbours. A common characteristic of many farmers is that they get their advice from neither the extension officer nor the man in the shop; they copy their neighbours. If they see their neighbour spraying, they will spray, just in case he or she gets ahead. But if they see their neighbour not spraying, perhaps that will work as well.

Kenmore: You also have to have the policy change. There is one major hurdle to progress in rice IPM in the next five years. 50% of the pesticide used on rice in the world is used in Japan, which grows only 2% of the rice. This is because the support price for rice in Japan fluctuates at between six and nine times the international trade price. In January 1992, Watanabe, the Foreign Minister, announced that the planned response of the government of Japan to the GATT negotiations was to shift from a complete ban on the import of rice in Japan to a 600% tariff. This sounds like a staggeringly large figure, but that's exactly what the support price is for rice in Japan. The trick is that they are going to slide that tariff from 600% to 15% over 5–7 years. As that happens, roughly 40–45% of the world's market for rice pesticides will evaporate. The world's market for rice pesticides is over US $ 2.5 thousand million. There is going to be tremendous pressure from industry; every chemical company in the world sells something in Japan. There is going to be tremendous pressure on governments in rice-growing Asia to absorb that evaporated market. These countries don't need any of that pesticide—they have to reduce their current level of rice pesticide use. It's going to be a tremendous challenge to policy-makers in India and in Asia generally to put up effective barriers. Tariffs are probably not the answer, because it's very hard to keep them going. You will probably need non-tariff barriers, which are allowed for pesticides under the London Guidelines of the United Nations Environmental Program of 1987 on international trade in potentially toxic substances. These include such famous stars as plutonium; the same provisions apply to pesticides. Policy-makers are going to have to use those sanctions to regulate the predatory tactics that are going to be used to get a hold of your market. Policy is going to have to protect IPM on that level. If and when that happens, as we work slowly in the village areas, we will have enough examples of villages based on the kind of group work and employment opportunities Dr Swaminathan has illustrated, to give governments a positive alternative. But the next 5–7 years are going to be critical for that reason.

Swaminathan: In relation to the transition from environmentally unfriendly to friendly techniques, there are several possibilities in India. For example, in the state of Assam, there is hardly any pesticide consumption and they use scarcely 5–6 kg of nitrogen per ha in terms of mineral fertilizers. An organization like the Indian Council of Agricultural Research should ensure that these areas do not follow the old pathway of stimulating farmers to use more fertilizers and chemical pesticides. There is a great opportunity to map these areas and see how we can increase productivity by more environmentally benign technologies. In Assam, nitrogen-fixing green manure crops can be grown extensively before rice, because soil moisture is not a limiting factor. Similar opportunities for introducing immediately eco-friendly technologies should be identified.

Kenmore: The experience recently is pretty scary. I was in Cambodia last week, which everyone says is in the same low pesticide use situation as Assam. I went by land from Ho Chi Minh City across the border to Phnom Penh. I stopped at every market I saw. In every market, you could buy three or four classic pesticides, such as methaparathion, metamidaphos and monocrotophos. They were usually being displayed next to the cosmetics in the market stalls! I agree we want to build on the potential of places like Assam which should be at a very low pesticide status right now, but we have to be realistic about what's really flowing in there across the borders.

Summary

Jan C. Zadoks

Department of Phytopathology, Wageningen Agricultural University, PO Box 8025, NL-6700 EE Wageningen, The Netherlands

When the hour approaches to say farewell, we tend to become philosophical. Once upon a time, mankind believed in magic—the prescientific or magical paradigm. During the 19th century, a scientific battle was fought, very fiercely, and a new paradigm came into being (Zadoks 1991). Scientists, the public and politicians accepted that a pathogen was the cause not the sequel of a disease. In around 1890, that was a new idea. Thinking about disease changed radically with the advent of antibiotics after the discovery of penicillin in 1940. Prevention and precaution were no longer necessary, because 'a pill cures any ailment'— the new chemical paradigm. This was also felt in agriculture: farmers treat a crop literally to relieve anxiety. I call this the sleeping pill treatment.

The enlightened people who, for ethical, philosophical, religious or scientific reasons wanted something different, had a very difficult time. Another battle was fought on philosophical and scientific fronts, and a new paradigm emerged—the environmental or sustainability paradigm. We can also extrapolate and call it the solidarity paradigm, referring to international solidarity which is now greatly lacking. If the Uruguay round of the GATT (General Agreement on Tariffs and Trade) negotiations succeeds and the European Community joins in, this will be a great advantage to the developing countries.

Now we look at things with new eyes. As of about 1990, we have put on new spectacles to see more clearly. Elements of our new vision include the ecological viewpoint, the long-term view, the transition view, the tritrophic systems, good seed supply by breeding and seed treatment, field monitoring, forecasting and farmer field schools. The need to re-educate the masses to give them self-confidence (to help them keep their feet in the information war, as one Javanese farmer phrased it recently), has brought natural scientists and social scientists together—such was the design of this meeting, and it worked.

Although we may not agree with each other in all details, we did experience a feeling of togetherness. The take-home message is that the two cultures need each other, complement each other, and strengthen each other. We are working for empowerment of the farmer. Such empowerment will have far-reaching consequences. Farmers as decision-makers with self-confidence in technical matters take pride in their knowledge. They become different citizens—proud and vociferous. Empowerment is a democratizing process; it strengthens democracy. Sustainability requires institutional change in governmental, educational and industrial systems. This was illustrated in the paper by Swaminathan and the following discussion. With sustainability may come equity.

We spoke of two cultures, natural and social scientists; two other cultures are men and women. The need to 'meet a farmer and her husband' is not the idea of a lunatic sociologist; it is the field experience of many natural scientists working for sustainable crop protection. The sex issue is an urgent one. In the foreseeable future, women farmers will carry more of the burden, not less. Trans-generational sustainability of agroecosystems is very much women's affair. They bear, feed and educate the next generation; because of this, women farmers should be approached in an appropriate way, different from the approaches used until now. Besides, in The Netherlands and in Indonesia, women usually hold the purse-strings and co-determine the purchase of pesticides. They have their own ways of conveying a message to their husbands. These and many other things we have learned at this Ciba Foundation symposium.

We also learned that Asia is leading the way rather than following towards sustainability. It has even developed a yardstick to measure progress along that way. We are very impressed by the progress made in our host country, India, and specifically in the host state, Tamil Nadu. We are also impressed by the strength of the NGOs operating here, small and large ones.

Ladies and gentlemen, thank you once more for your contributions. Fare well.

Reference

Zadoks JC 1991 A hundred and more years of plant protection in the Netherlands. Neth J Plant Pathol 97:3–24

Index of contributors

Non-participating co-authors are indicated by asterisks. Entries in bold type indicate papers; other entries refer to discussion contributions.

Indexes compiled by Liza Weinkove

Subject index